TECHNOLOGY AND TRANSCENDENCE

Technology and Transcendence

Edited by

Michael Breen
Eamonn Conway
Barry McMillan

First published in 2003 by
the columba press
55A Spruce Avenue, Stillorgan Industrial Park,
Blackrock, Co Dublin
www.columba.ie

Cover by Anú Design
Origination by The Columba Press
Printed in Ireland by ColourBooks Ltd, Dublin

ISBN 1 85607 384 X

Contents

Introduction

We live in a society in which modern technology is all pervasive. We experience the various ring-tones of mobile phones at business meetings, concerts, on the bus and in a myriad of other places. Domestic satellite dishes are now a normal sight in our towns and cities. The Internet is taken for granted. Children expect to have access to personal computers and are increasingly expected to be computer literate at an early stage of development. People of all ages are commonly seen in public with earphones listening to MP3 files whose content varies from Bach to Blur. Instantaneous reporting of world events has become an expected norm, to such an extent that the many could sit back and participate in the invasion of Iraq from the comfort of their living rooms, experiencing some of the war's horrors without the attendant stench of death or an awareness of bloody aftermath.

Alongside such developments various caveats have been sounded. The spectre of Orwell's big brother has come home to us anew with national databases holding details of our salaries, taxation, health, and insurance arrangements. Our footprints across the electronic desert of the Internet can be readily tracked both by website vendors and our own Internet service provider. E-mail, regarded by many as secure, is simply an electronic postcard, the contents of which can be readily seen by those who have the technical 'know how'. The Internet itself is sometimes represented as the devil's playground, wherein children are stalked daily by paedophiles intent on destruction and mayhem.

But the benefits of technology cannot be denied. Technological advances in medicine, for example, allow us the seeming miracle of MRI scans and telemedicine, with robots able to perform certain surgery while guided by a surgeon on a different continent. The new communications technologies and the digital revolution are generally hailed as 'a good thing', far beyond the Sellers and Yeatman comic sense of that phrase.

Society is swept up in technological development whether it wishes it or not; some banks, for example, are compelling their customers to use electronic swipe cards for all transactions. The digital revolution, then, is here to stay, and technology continues to impact on all our lives, from the incubator to the crematorium with a great many technological experiences between the two.

This book represents the work of nineteen scholars in four disciplines: philosophy, theology, sociology and cultural studies. Drawing on the work of Martin Heidegger, Jones Irwin explores the modern understanding of technology as 'mere means' in the context of a more holistic understanding of technology as 'a way of revealing'. Fiachra Long takes issue with Heidegger's approach, rejecting it as 'hardly good news'. John Sharry and Gary Mc Darby present concrete evidence of the positive application of technology, drawing in particular upon their experience at Media Lab Europe. They illustrate their point well with their opening anecdote: are we 'heading the wrong way' as the philosopher states, or should we believe the assertion of the engineer and the scientist that we are in fact 'making great progress'? The real question is, of course, progress towards what precisely? Stephen Butler Murray asks some leading questions with regard to the nature of technological progress, while Jim Corkery writes about what he understands as a clash of logics, the I-centred logic of the technology paradigm as opposed to the other-centred logic of the Christian paradigm. Brian Donnelly develops another dimension of this clash, by examining the work of Paul Tillich who regarded 'the ascendancy of technical reasoning as a devastating assault on reasoning itself'. Rik Van Nieuwenhove continues the theological debate, arguing that 'technology should not become an end in itself, nor should it be rejected in a utopian vein, but should be seen for what it is: an instrument to be put to a proper use.' Such a statement inevitably begs the questions: what is the 'proper' use of technology, and who gets to decide?

Paul Downes returns to the philosophical concerns, citing Foucault's understanding of 'post-modern truth claims ... based not primarily on reason but on power'. Cardinal Paul Poupard suggests that technology is not something of which we should be afraid, but that the human mind can exercise wisdom in the creation and adoption of new technologies, as long as the idea of the human person is kept to the fore. While acknowledging concerns about the global technological revolution, Mark Dooley argues that it is technology itself that holds the greatest promise to the poor, in far greater proportion than any risk. Paul Brian Campbell, on the other hand, presents a compelling case of failure in the brave new cyber world to deliver the much vaunted promise of prosperity, so much so that the world is now riven by a digital divide as a further separation between the haves and have-nots.

Barry McMillan takes up the reflection by means of a critical perspective on cinema, and through examination of the science-fiction film *The Matrix* incisively illustrates dominating trends in contemporary culture. Mary Corcoran *et al* approach the topic of technology from a very different point of view, examining the impact of transportation and telecommunication in a new sub-urban community, with a particular emphasis on spatial proximity in relation to technology. Michael Breen looks at the empirical evidence on attitudes to technology from the European Values Surveys and then considers some of the critical-cultural issues that arise from such an empirical perspective. He concludes that 'despite the shrinking of our world by means of technology, we have forged a whole series of individual little worlds, rather than a single open communicating world of equals, a globe of villages rather than a global village.'

Stijn van den Bossche reflects on the film *Dekalog I*, which in his view is a clever deconstruction of modern technology as idol. Finally, Eamonn Conway completes the series of reflections with a chapter dealing with Christian anthropology, in which he concludes that 'The goal for the Christian community should be the construction of the kind of free space and silence … in which shared reflection with technologists and scientists can take place …'

These essays are intended to contribute to the creation of a space for shared reflection. In keeping with the mission statement of the Centre for Culture, Technology and Values, Mary Immaculate College, University of Limerick, they are intended to stimulate critical reflection regarding the impact of technology on culture and values in light of the Christian understanding of human dignity and social responsibility. It is our hope that they do so.

Michael Breen
Eamonn Conway
Barry McMillan

CHAPTER 1

Does Technology Squeeze Out Transcendence – Or What?

Jim Corkery

Introduction

Technology, it is said, plays havoc with transcendence, turning human desires for the infinite into finite 'satisfiables' that can be delivered by techno-wizardry. Thus technology flattens out the human landscape into an arena peopled essentially by producers and consumers. Technology is accused of easing the Transcendent Other – the one Christians call 'God' – out of everyday life; after all, as Bertrand Russell is quoted as saying, people in sailboats have far less difficulty believing in God than people in motor boats do (Kilcoyne, 1997). Technology seems to promise the earth; but that's all! The question arises: is what it 'seems' to do all that it actually does, or does the – admittedly dazzling – horizontality of the techno-world nevertheless leave space for some vertical light, some outreach to (and 'inreach' of) the transcendent? Does technology simply corral human activities and values into the straitjackets of 'use' and 'enjoyment'? Or is it possible that there can be something more in the picture?

People in an Age of Technology – Some Trends and Examples

Where technology is concerned, the emphasis is on know-how and on efficiency. The idea is to do things competently and swiftly. The focus is on production, on getting things done, on achieving results. Whether the technology is industrial (an electronic engineer designing a robot that can paint a thousand cars in an hour); or medical (a research scientist developing new drugs in a laboratory); or communications-oriented (a computer scientist producing a clamshell mobile phone with a colour display built-in digital camera), the goal is efficient, high-performance, profitable production. The notions of building bigger and better, faster and finer, are dominant. The main goal is: production for consumption for profit. All else is essentially subordinate to this goal, even when what is produced could be said to enhance the quality of human life. In this scenario human reason tends to be instrumental (use-oriented); human activity tends to be productive, productive-for-consumption; and human beings tend to be seen as consumers, who exist to buy, indeed who *are* to the degree that they *do buy*: *tesco, ergo sum* (Grey: 1997, 1). Gradually a way of living (a culture, indeed) is generated in which instrumental reason, technical skills and conspicuous consumption occupy the highest places. Aesthetic, imaginative, poetic, speculative, artistic and religious thought all remain lower on the totem-pole than instrumental, productive reasoning. Sheer technical ability – to make, produce, construct – takes precedence over those distinctively human activities that have their ends in themselves, such as, celebrating a meal; parenting; worshipping; making a home; working at a marriage. In the final analysis, it is the *doing* that defines us – not 'who are you?', but 'what do you do?'

The epitome of these circumstances is the *self-made man* (I use the sexist word consciously): the essentially unrelated person who owes nothing to anyone, who has made it to the top by effort and skill, who is self-reliant (has insurance for everything – from dental health to bad weather on vacation) and can stand on his own. Such a one is autonomous, celebrates his autonomy and is rewarded for it by society, which grants him access to all the material comforts (houses, cars, holidays), to the fashionable places (clubs, chic parties, cultural events), to unlimited credit and to much public approbation. There is little space for thankfulness in this man's life because what he is, and who he will become, and how he will shape the world are all something he must achieve by and for himself (Greshake: 1977, 13). Other persons in the self-made man's life are numerous – success breeds popularity – but these other people tend to be what might be termed 'contacts' rather than friends, people who can be made use of, or patronised, rather than friends to be enjoyed, or helped, or simply loved. 'You're off to Frankfurt,' said one of the self-made to me recently; 'I've a contact there, just mention my name, let me give you his number' (and presto, the pocket-organiser is produced and spews forth the name of the electronically-stored genie who can be counted on to make my life better in Frankfurt because my 'self-made man' had made his better some time ago). Who is this contact in Germany? Basically, someone whose *doing* takes precedence over his *being* – someone who, in the end, is valuable because usable.

Twenty years ago, when computers first came into schools, I noticed that when the students learned programming, they quickly became fascinated by the ways in which they could get the computer to respond to them as they wished, depending on how they programmed it. Thus words of praise, affection, admonition, humour, etc., would pop up for them, depending on how they had set up the computer to react to their different inputs in the first place. The computer became person-like for them – it even had a name: Jessica – and so you would hear students speak about what Jessica 'said' to them that morning or 'replied' to them that afternoon, as if Jessica had 'a mind of her own'. Her (I mean its) responses were in fact nothing more than the product of her dialogue-partners. They were always the ones in control. And now they could not be surprised. This is fine with computers, perhaps, but when it becomes a skill that is transported into the interpersonal sphere, then relationships – and the very ability to relate – can genuinely suffer. We have come a long way since twenty years ago. And we have seen many young people relate far more to their computer screens (the new tabernacles) than to their families and classmates during that time. The opposite is true too, of course; and many an e-mail helps to keep friendships alive. But there is nothing automatic about this; and e-mail itself can have quite a shadow-side as people chat endlessly with strangers in cyberspace while not knowing who their neighbours are next door.

Culture in an Age of Technology – Some Trends and Characteristics

The kinds of behaviours I have been talking about – those of the self-made man or of the newly-dazzled computer geek – are embedded in and spring from a cultural matrix that has been in large part produced by technology and now influences, insidiously and *sotto voce*, the realm of human relationships and human values. Matthew Lamb speaks of the modern age – surely the age of technology – as characterised by a 'master-metaphor' of movement, of action-reaction, that pervades the sphere of human relations and values as a kind of mentality of: act-or-be-acted-upon (Lamb: 1987, 785). Marx (and others) lie at the root of this mentality, which divides the world into oppressors (actors) and victims (acted upon) and thus promotes a sort of conflictual understanding of the human world as a sphere of 'eat-or-get-eaten'. The twin team here is simply activity / passivity: do or be done unto. And you can see it played out in the above-described relations of the self-made man and even of those 'net freaks' who live mostly in a largely relationless, manipulable and manipulative, virtual world that aggressively bombards them with seductive and exploitative offers and sucks them into shallow relational spaces where often the most intimate things can be said but only because no real relationships can ever develop from them. In all of what happens here, an aggressive 'act-or-be-acted-upon' culture is being established in which the truncated transcendence of men and women is taken advantage of as they are offered products for their consumption that in reality consume them.

Without being determinist, it is evident that people can be harmed in circumstances such as these. They are being shut down, closed off at the top, assured that they can be satisfied by inventions, contraptions, packaged experiences, tinsel-things. The surprise, unpredictability, unmanipulability, freshness, care for persons-as-persons is being taken out of their relationships and these relationships are becoming functional: a means to something rather than an end in themselves. If the 'something' does not happen, the relationships are quickly questioned – and often abandoned in this culture of non-support. The phenomenon is not just individual; the patterns that characterise these relationships are being 'writ large', inflated into the very ways in which social organisations and social life are constructed. Thus the technical-rational mentality is becoming institutionalised, with businesses, offices, restaurants, and even churches and homes being 'rationalised', that is, organised in ways that make the achievement of goals and the smooth 'processing' of people primary. This is visible in airports, shopping centres, financial institutions, healthcare facilities, educational establishments (to name but a few) and – yes – in churches, and even in families (where the microwave has ensured that people need not eat together any more). No one can deny the splurge in paperwork, record-keeping and quantifying that has mushroomed in recent years. In academies, the measuring, recording, documenting and processing is eternally evident and evidently eternal! The characteristics that have been put forward as the hallmarks of 'McDonaldisation' (Drane: 2000; Ritzer: 2000), efficiency, calculability, predictability and control, are ever more in evidence in our dealings with one another – in social groups as small in size as the family and as large as government departments and multinationals. The fast food restaurant is the measure of all things human: processes must be quick, predictable, quantifiable and – above all – controlled. In all of this persons, and the value attached to them, and the values pursued in relating to them, are subsumed under the culture of the fast-food restaurant, where there is just about enough to keep people alive and fed in the short-term, but definitely not enough to nourish and maintain them in health in the long-term.

What I have been describing is pretty rough stuff – certainly not a welcome scenario for people. Nor is it just a matter of technological development giving rise to a McDonaldised culture that devours the best in people; somehow, we go along with and support this. There is a complex, hen/egg type situation here, in which technology begets rationalised cultural patterns of behaviour, which in turn become institutionalised, acquiring a quasi-independent life of their own, such that, as culture, they act back on persons to further 'technologise' them and their relationships. The culture is in us and we are in the culture. We use terms such as 'society' and 'culture' to try to put words on that intangible, but very real *whole* that is more than the-sum-of-the-parts wherever human beings are gathered together and that shapes us even as we shape it. Insofar as the two terms can be distinguished – and this is not easy – 'society' might be said to refer to the more outward or visible dimensions of human togetherness: social structures, organisations, collectivities, and the like, whereas 'culture' might be said to refer to the more inward or invisible dimensions of human togetherness: ideas, attitudes, mind-sets, thought-patterns, shared beliefs and ideals, and so on – all of which, of course, will manifest themselves in visible behaviours such as customs, mores and practices, and even in visible 'constructions' such as academies, supermarkets, art museums, clubs, businesses, and even churches; and then we are back with institutions, and thus with 'society', once again! So one sees: 'culture' and 'society' are somewhat indistinguishable (Jenkins: 2002, 62-63) and are so central to what it is to be human that they might be described as the very water we swim in or the air we breathe. And yet, for all their unnoticeability, our lives depend on them as they both shape and express us, of whom they themselves are the very shape and expression.

According to Leonardo Boff, cultures can be understood as

embodying overall or collective projects akin to the personal life-projects of men and women (Boff: 1979, 141-147). If an individual (thus moral theology) has a fundamental option that more or less expresses the basic intent and direction of his/her life, so too, albeit analogously, a culture can be seen as embodying a fundamental option, a collective life-project. This is the case, even though cultures are not uniform and, in fact, contain a variety of sub-cultures. Notwithstanding this internal variety, one can still speak of this or that culture (Irish culture, even 'western' culture) as having a kind of overall physiognomy that more or less manifests the collective face of a particular people at a given time. Thus we speak of a culture as being more or less compassionate, more or less individualistic, more or less acquisitive, caring, or whatever. It is the overall project characterising a culture that people have in mind when they ascribe to it adjectives such as the ones I have just used. All cultures are ambiguous morally – in terms of the values they enshrine or embody. So far I have been highlighting the morally dubious values of technological culture(s): the ways in which they tend to reduce people to 'usables' and to make the world uni-dimensional by denying humanity's transcendent origins and destiny. However, as with persons, so too do all cultures live under the dialectic of sin and grace – embodying, that is, not only anti-transcendent, anti-divine dimensions, but also dimensions that foster human transcendence and open men and women out to one another and to God. I wish to search a little for evidence of, and possibilities for, such dimensions now.

The 'Openings' and Possibilities, the Realistic Hopes, Discernible in the Present Situation

With Leonardo Boff it has been put forward that cultures, like persons, embody fundamental options that both express the aggregate of personal decisions and also influence subsequent decisions of persons. But is it the case that a technological and highly rationalised culture will inevitably embody (and promote) merely destructive tendencies or are there any signs of hope and opportunities for rediscovering personal values evident in our present situation? I think so; and, in a search not unreminiscent of Peter L. Berger's uncovering of 'signals of transcendence' (Berger: 1969, 70f) over thirty years ago, some of them may be mentioned here. There is a hunger for meaning and a thirst for spirituality evident in present-day Irish culture that offers much ground for hope, even if not all the spiritualities embraced are free of anti-divine dimensions. There is a greater appreciation of, and search for, different forms of community. A willingness to get involved in issues that require public action such as protest and demonstration seems also on the rise (witness, for example, the actions of those who are protesting against Ireland's endorsement of US intent with regard to Iraq without its people being consulted). The uncertainty of the 'self' in contemporary, postmodern culture has begotten strange kinds of individualism, to be sure, but also given rise to new conversations between the non-like-minded that are far less arrogant and far more tolerant and open than anything that characterised Irish debates in the past. Certain kinds of media action have led to greater discovery of truth and a greater taking of responsibility for previous inappropriate actions. Furthermore, the dominance of the image, for all its ambiguity, offers a way of appealing to aspects of persons (affectivity, spiritual hunger, care of the earth) that modern, rationalised culture tended to reject. The voices of the unvoiced are heard more loudly now, following thirty years of liberation theologies, and, amid an increasing and evident plurality of religions, there is much greater openness to interreligious conversation and sharing. Coupled with all of this is the growing number of laypersons refusing to accept church life as it was – with themselves consigned to the margins – and many are coming forward to seek and to take a more responsible place in the life of the Christian community. All of these things – and there are more – can be seen today as concrete 'signals of transcendence' in what often seems to be a horizontal, entirely immanent existence. Some might argue that these signals have nothing to do with our technological culture as such and that they in fact represent something of Irish culture's move away from the harshness of the 'modern' to the greater openness and pluralism of the 'postmodern'. Indeed they do represent such a move, but why cannot that be a 'signal of transcendence', an incipiently-graced opening of culture to a kind of living in which technology, profit and goals no longer call all the shots? Indeed a theologian of grace might well argue that the presence and character of these concrete 'signals' – indicating possibilities where there is blockage and gently inviting breakthroughs to transcendence – are precisely grace emerging at the 'stuck-points' of human history in a shape divinely tailored for the healing and transformation of at least some elements of the current malaise.

Which brings us specifically to the Christian heritage and to the possibilities it might offer today. What resources might Christianity have for building on the opportunities just noted and for strengthening the values of inclusiveness, community and justice/peace that can be seen to underlie them? Even at first glance it can be seen that Christianity is characterised by a 'form' or 'shape' – a kind of 'logic' – that runs fundamentally counter to the 'form' or 'logic' of a technological culture that views people primarily as instruments for profit and that sees individual rights as limited only by the rights of other individuals but not at all by the social-justice responsibilities that are attendant upon all claiming of rights. The 'logic' of Christianity is ex-centric, self-emptying, other-directed; this is evident from the kenotic Christ-form that occupies the heart and centre of the Christian story. The 'logic' of the technological culture that we have been discussing is I-centred, focused on my rights and possibilities, basically ego-centric. The two 'logics' clearly clash. That of technological culture deems it sufficient to agree that the rights of individuals, who 'self-make' and have every entitlement to do so, are circumscribed only by the rights, considered abstractly, of other individuals, but not by the concrete circumstances of those individuals that may well be denying them equality-of-access to those rights. The 'form' or 'logic' of a mentality such as this is in sharp contrast to the 'logic' that is prophetically evident in the kenotic Christian form, the form of self-emptying (the ec-centric form) that is revealed to be at the heart of the very being of Jesus Christ – and hence of God – in Christian revelation. This Christ form, then, as a resource from Christianity, is a positive challenge to the anti-divine aspects of contemporary ego-centric cultural forms and offers the opportunity of envisaging ways in which people can live together that embody values very different from those of use and function. Theologian David Schindler has pointed to the gift at the heart of Christian revelation that the kenotic Christ-form is for those who would wish to shape a cultural project not against, but in the direction of, the divine intent for humanity (Schindler: 1990, 21); and John Haught has indicated how the letting go and letting be of the self-emptying Christ makes visible the way for dialogue with men and women of other religions (Haught: 1993, 80-82) and, I would add, for dialogue with all those persons who are in any sense a challenge to one's own self-centredness, individual or cultural.

Theologian Matthew Lamb (mentioned earlier) has drawn attention to the rediscovery of a contemporary (postmodern) meaning for the notion of 'praxis' that returns it to its classical roots and steers it away from its association (even in Marx) with an 'action-reaction', an 'act-or-be-acted-upon' mentality, that tends to turn persons into usables and their activities into merely instrumental or goal-directed undertakings. Aristotle's vocabulary admitted of a distinguishing between theory, praxis and practice, the first referring to theoretical thought, the third referring to productive activities (with goals outside of themselves) and the second referring to specifically human operations that respected the subjectivity and freedom of persons and gave rise to behaviours and actions which were undertaken and valued for their own sakes. The productive activities are exemplified in such actions as: cooking a meal, building a house, designing a university campus, preparing a liturgical space, constructing a city. The activities that are praxes are exemplified in such undertakings as: celebrating a meal together, making a home, creating a community of learning, participating in a liturgy, developing a civic community. A conscious recovery of such activities at the heart of families, of churches, and indeed in the wider institutional life of a community is essential if people are to be treated as persons, not numbers; if they are to be met, not simply processed. Think of the revolutionary change that such a recovery would bring to the life of the church, which has become so formal and bureaucratised: more and more characterised by the hallmarks of McDonaldisation. Think of the ways in which refugees and asylum-seekers might expect to be dealt with if 'praxis' rather than mere 'practice' were to be the emphasis in our human relations. And if you have any doubt as to how such praxis looks, just glance at the dialogues between Jesus and Nicodemus, the woman at the well, the woman taken in adultery, the man born blind and Martha in chapters 3, 4, 8, 9 and 11 respectively of John's gospel. The methodology, the 'praxis', is there. If introduced, it could cause a revolution.

People are seeing now – in the case of the church, for example – that it is less the *what* than the *how* that is wrong. The *how* has become so bureaucratic, so rationalised, so impersonal, so cold. People are met as quantities, as consumers, even as victims (when these are met); but they are not often met as persons. They are handled and processed, sent to the relevant departments, managed, controlled – but not greatly appreciated, listened to, or loved. One might say: we have dialogues now. Yes, endless ones. But the reports of the dialogues end up on shelves because the rationalised, McDonaldised system wins. Technology triumphs, or so it seems. Yet there is hope here too. I will end with a word from John O'Malley, who recently pointed out that the Second Vatican Council was much more about the how of the church than about its what (O'Malley: 2003). This how emphasised collegiality, shared ministry, religious freedom, attention to the signs of the times and to the real griefs and anxieties of people. It had to do with how we were to deal with people, and in particular with those who differed from ourselves. In this it was prophetic, having the kenotic 'Christ-form' that 'lets go' and 'lets be' in mind long before the theologians I mentioned began to even see it and the possibilities it offers to this troubled, yet salvageable, technological age.

Bibliography

Berger, Peter L. (1969), *A Rumour of Angels: Modern Society and the Rediscovery of the Supernatural*, (Garden City, NY: Penguin).

Boff, Leonardo (1979), *Liberating Grace*, (Maryknoll, NY: Orbis Books).

Drane, John (2000), *The McDonaldization of the Church*, (London: Darton, Longman & Todd).

Greshake, Gisbert (1977), *Geschenkte Freiheit: Einführung in die Gnadenlehre*, (Freiburg-im-Breisgau: Herder).

Grey, Mary C. (1997), *Prophecy and Mysticism: The Heart of the Postmodern Church*, (Edinburgh: T&T Clark).

Haught, John F. (1993), *Mystery and Promise: A Theology of Revelation*, (Collegeville, MN: The Liturgical Press).

Jenkins, Richard (2002), *Foundations of Sociology*, (UK: Palgrave Macmillan).

Kilcoyne, Colm (March 9th 1997), 'A hated God is better than no God', in *The Sunday Tribune*, (Dublin: Ireland).

Lamb, Matthew L. (1987), 'Praxis' in, Komonchak, Joseph A., Collins, Mary, and Lane, Dermot A., (eds.), *The New Dictionary of Theology*, (Dublin: Gill and Macmillan).

O'Malley, John W., SJ, (2003), 'The Relevance of Vatican II',

(Cambridge, MA: 2003 President's Letter, Weston Jesuit School of Theology).

Ritzer, George (2000), *The McDonaldization of Society*, (London: Pine Forge Press).

Schindler, David L. (1990), 'Introduction: Grace and the Form of Nature and Culture' in, Schindler, David L., (ed.), *Catholicism and Secularization in America*, (Indiana: Our Sunday Visitor).

CHAPTER TWO

The Future Is Now: *The Matrix* as Cultural Mirror

Barry McMillan

Introduction

In 1984 William Gibson's novel *Neuromancer* – the story of a burnt-out software pirate forced to do one last job in Earth's computer matrix – was published, and the English language gained a new word: cyberspace. With the contemporaneous surge in the uptake of internet access and usage, 'cyberspace' rapidly became lexically commonplace and entered the *Oxford English Dictionary*, whilst, as a genre, 'cyberpunk' spawned numerous Gibson copyists, in book, graphic novel, film and animation forms. Gibson gained a pre-eminent status in the genre as a result of *Neuromancer* and subsequent books such as *Mona Lisa Overdrive,* and *Idoru,* and became dubbed 'the godfather of cyberpunk'. But, like many before him who have created art notionally about the future, Gibson revealed in an interview earlier this year:

> The thing is, I'm not actually writing about the future. What I am doing is using all these great things that came in my science-fiction writer's tool kit to take snapshots of the unthinkable present. (Mallon: 2003, 26)

Gibson's comment thus confirms the view that to a greater or lesser extent, those who create art about the future reveal more about the present out of which they extrapolate, than the future which they posit. Between 1946 and 1949 George Orwell wrote *Nineteen Eighty-Four*, and provided greater insight into, and description of, the period leading up to the book's conception than of the year of it's setting. Similarly, Aldous Huxley's *Brave New World* published in 1931 is infused with the malign influences of a totalitarian Soviet Union and fascist Italy. Stanley Kubrick's *2001: A Space Odyssey* (1968) is redolent with the wide-eyed aspirations Americans had invested in the space programme of the late-1960s.

In the light of such examples, it is my working premise herein that, whatever prescience futurist art may contain, it more significantly offers up telling insights about the culture within which it is created. Such art – sometimes deliberately, sometimes inadvertently – (re)presents a distillation of significant trends widely diffused within the culture of its origin. In addition, therefore, to whatever other purposes, artistic or otherwise, it might serve, such art provides a snapshot of its *zeitgeist*, and reflexively, acts as a mirror of the culture that produced it. It is in this light then, that I turn to examine the 1999 feature film *The Matrix*.

The Phenomenon of The Matrix

It might well be asked why *The Matrix*, rather than some other example of popular culture, is called on to bear this scrutiny, and indeed, it is the case that other examples might well be fruitfully examined. With regard to the choice of *The Matrix*, however, the answer is straightforward. *The Matrix* is a (pop-)cultural phenomenon with few equals.

From a cinematographic point of view its visual stylings genuinely added something new to the visual grammar of cinema, and to such an extent that four years after the film's release it is still being 'quoted' in numerous other films, films as diverse as *Crouching Tiger, Hidden Dragon* (2000), *Shrek* (2001), and *Equilibrium* (2003). Its stylings have, in addition, permeated a variety of other pop-cultural settings, most noticeably fashion, electronics (through tie-in sponsorship deals), and marketing. At the time of writing, for example, a television advertisement for a newly-launched breakfast cereal is using *The Matrix*'s 'bullet-time' 360º visual swooping technique to convince consumers of the merits of eating bran first thing in the morning. That the marketing of fibrous roughage is via the means of a cinematic styling developed to visualise virtual reality, is indicative of *The Matrix*'s remarkable degree of cultural embeddedness and ubiquity.

Furthermore, on its release, *The Matrix* generated more than $170 million in box office ticket sales in the US alone, and the DVD subsequently became the first to attain sales in excess of one million copies. It has generated two sequels, *The Matrix Reloaded* and *The Matrix Revolutions* – both to be released during 2003 – and at the time of writing, the first of these has generated ticket-sales of almost $136 million in its first four days of release. The release of *The Matrix Reloaded* was accompanied by the simultaneous release of *The Animatrix* – a DVD of nine related animated stories – and *Enter The Matrix* – a computer game providing the backstory for one of the characters in the new film. Such impactful extension across multiple media affords *The Matrix* phenomenon a particular significance.

However, as well as being a cinematic, marketing and (pop-) cultural phenomenon, what additionally marks out *The Matrix* is the seriousness with which it has been dealt by some within the academic community. The films' official website www.whatisthematrix.com features a philosophy sub-section in which academics from Australia, the UK and the US – including Berkeley's Hubert Dreyfus – reflect on the content of the film and the questions it raises. Cornel West, the Princeton professor, has gone one step further and performs a role as a Council-member in *The Matrix Reloaded*. On 9 June 2003, the Vatican-endorsed online newsagency Zenit.org carried an interview in which Juan José Muñoz Garcia, a Spanish professor of anthropology and ethics, outlined his belief that *The Matrix* was a 'lesson in anthropology', and that, '… the main character … lets us observe how truth and ethics go hand in hand.' (www.zenit.org: 9 June 2003)

Numerous books are available, not only on *The Matrix* and science fiction as might be expected, but on *The Matrix* and science, *The Matrix* and spirituality, and *The Matrix* and philosophy. It is in the light of both such widespread, and such particularised, endorsement, that I believe that *The Matrix*, and what it reveals about our culture, warrants examination.

The Matrix

As a result of the impact of *Jaws* (1975) and *Star Wars* (1977), Steven Spielberg and George Lucas, respectively, are credited with the creation of the contemporary blockbuster movie. Both films were characterised by their simple narrative storytelling and their primary reliance on visceral thrills. Their unexpected huge international popularity provided Hollywood producers with the template for the big-budgeted, widely-marketed, high-impact, blockbuster movie now familiar to cinema-goers. Recent examples include *Gladiator* (2000), *Lara Croft Tomb Raider* (2001), and *Spider-man* (2002). As is clear from these examples, the dominant characteristic of the contemporary blockbuster is the 'roller-coaster' thrills it provides. *The Matrix*, whilst clearly belonging to, and marketed within, this adrenalin-inducing genre, also expands the boundaries of the genre, and more significantly goes uncharacteristically against the grain in noteworthy ways. It is this expansion, and apparent contradiction, of genre conventions that has enabled it to gain credibility in circles beyond the usual blockbuster audience.

Though not explicitly sexual in its narrative, *The Matrix* is a lesson in seduction. Distinguished by a kinetic visual style both arresting and exhilarating, a production design that all but pulsates with a glamorous hyper-reality, and costuming that magnetically attracts the eye and generates a palpable frisson through its fetishisation of black designer clothing and dark glasses, the film draws the viewer in through an excess of stimulating visual sensation. It is all but impossible to resist its allure. Thus, in the film's excess lies much of its success, for the sensory overload and the adrenal-endorphic cocktail this overload induces have the capacity to saturate the viewer's critical faculties and leave them neutralised.

Whilst its visual qualities distinguish *The Matrix* from others of its genre, so too does the fact that its plot is complex and infused with philosophical, mythological, and theological concerns. Created and maintained by rogue computers, the Matrix of the title, is a computer-generated collective delusion of late twentieth-century reality, hard-wired into the minds of an enslaved humanity, unknowingly stacked in vats in vast body farms, where their limp bodies passively generate bio-energy that maintains the computers. The narrative of the film follows the small group of humans who have managed to break out of the Matrix, in their battle against it, and in their struggle to free the rest of dormant unsuspecting humanity.

Thematically the film touches on questions of free-will and destiny, the nature of reality and perception, technological development and its consequences, fate, and hope. That it also makes visual reference to Baudrillard's *Simulacra and Simulation,* and alludes to the plots of *Alice's Adventures in Wonderland* and *The Wizard of Oz,* is peopled with characters called Trinity, Neo, and Morpheus, involves a 'hover-craft' called the Nebuchadnezzar, requires that the characters make a visit to 'the Oracle', necessitates the salvation of a city called Zion, and centres on the hope for a messiah who will be 'the One', is illustrative of the varied complex sources on which it draws, and the serious and thought-provoking concerns it apparently aspires to deal with. *The Matrix* is therefore something of a paradox – a blockbuster that the viewer has to think about.

However, a film about the untrustworthiness of how things seem, *The Matrix*'s emphasis on surface appearance is instructive, and the key to discerning what truly lies at its heart.

The Matrix *as Cultural Mirror*

The Matrix is a film run through with deceit. Despite appearances, it is in obfuscation, and intellectual and ethical sleight-of-hand that the film's true virtuosities lie. Ostensibly a theo-humanist film about freedom, hope, and the dignity of human nature, if viewers can manage to pull themselves away from the massaging pleasure of immersion in the film's heightened imagery and aesthetics, what is revealed is distinctly discomforting, and the realisation of the extent of the film's powers of misdirection and seduction, unnerving.

I. Violence in *The Matrix*

Tank: 'So what do you need? Besides a miracle.'

Neo: 'Guns. Lots of guns.'

This exchange marks a key moment in the course of the film. It is the still point around which the film revolves, and the moment when what is in the film's soul is revealed. It is the pivotal moment to which the film builds up, and the sequence which precipitates the film's climax. It belies all the film's apparent philosophical and theological aspirations. Neo, the 'saviour', will initiate the new epoch with guns, lots of guns. For all the quasi-complex discussions about reality, free-will, fate, and freedom refracted across the surface of the film, this film is, at heart, an icy hymn to the perverse euphoria of unrestrained violence. Beneath *The Matrix*'s veneer of philosophy and mysticism lies an untrammelled indulgence of the Nietzschean will-to-power. Force is not only honoured here, it is rejoiced in and celebrated.

The images and action, choreographed to the last dropping dance of the last spent machine-gun cartridge bouncing on an impossibly gleaming marble surface, are magnificent in their effect and achievement. In the slowest of slow-motions (a new term 'bullet-time' was coined to describe the effect) havoc is wreaked minute upon minute, every second of the frenzy eliciting primal responses in the viewer. Never has force looked so appealing. Never has violence appeared so exultant. As Neo and

Trinity enthrallingly swoop and sweep in their omnipotent slow-motion through the mayhem they are realising, the depiction can only be described as balletic. Balletic violence? It is the obscenity of the conflation that startles and shocks and breaks the spell.

It is this film's greatest mendacity to make violence beautiful, and thereby attractive and desirable. It is unquestionably thrilling to watch the highly-stylised, slow-motion gunfights and hand-to-hand martial arts combat depicted, so breathtaking that the viewer is drawn to revel in such exquisitely wrought aggression – in thrall to the elation that watching it induces. But evocation of such response is a mark of the film's perversity, for what is created is a pornography of violence.

There can be no claim to outsidership for the viewer here. Distancing is not possible, for to justify engagement with the thrill of the violence the viewer must collude in the film's great narrative and ethical inconsistency. In its early sequences, the film goes to some length to explain the dynamics of its universe. The humans unknowingly and unconsciously trapped in their vats, experience a three-dimensional self – a self which is, unknown to them, a projection in the mass illusion which is the Matrix. Though this reality is illusory, 'physical' activity which occurs in the Matrix reality has genuine physical effects on the actual bodies of those hard-wired into it. Violence perpetrated in the Matrix has real consequences. People who die in the Matrix 'reality' die in actual reality. The humans in the Matrix are simply as our hero was before he was freed. The security guards therefore that are machine-gunned by Neo and Trinity at the beginning of the climactic sequence in the film, are therefore entirely innocent – humans projected into the Matrix reality in that role – and moreover, are the very people the rebels purport to be setting out to liberate. In the attainment of their cause, the rebel 'heroes' therefore heedlessly engage in carnage of the innocent. Human freedom, supposedly the film's great totem, is to be won with indiscriminate slaughter. The realisation is jolting. Behind the illusory theo-humanism of *The Matrix* there lies a harsh, repellent, totalitarianism, and to enjoy the film viewers are required to collude in it.

To recognise that there is in the film the attempt to make violence amoral, that there is a desire to elevate destruction to the level of the poetic, is to grasp that the film is steeped in a

Nietzschean ethic counterpointed with a fascist aesthetic. The film's fetishisation of costume suddenly assumes particularly disturbing meaning. In such a light, a new sensitivity to the film's true values emerges in the viewer.

II. Drugs in *The Matrix*

In the foregoing, I have made reference to the enticing immersive qualities of the visual experience of *The Matrix*. This aspect of the film takes on a more troubling note with the awareness that the film is littered with both implicit and explicit references to hallucinogenic drug-taking. In an early scene, Neo, warrantably confused by his computer's apparently spontaneous suggestion that he 'Follow the white rabbit', has the following exchange with Choi, a client to whom he is selling illegal computer disks:

> Neo: 'Ever have that feeling where you're not sure if you're awake or still dreaming?'
>
> Choi: 'All the time. Its called mescaline. It's the only way to fly.' [smiles]

Choi then knowingly suggests that Neo needs to 'get a little R&R'.

To the literary-minded the reference to the white rabbit preceding this exchange is clearly a reference to *Alice's Adventures In Wonderland*, and it does indeed alert the viewer to something of the spirit of disorientation of what is to unfold through Neo's subsequent discovery of the existence of the Matrix. However, it also serves a more covert purpose, in that it is also – to the initiated – an explicit drug-taking reference. It is long-suggested, though unsubstantiated, that Lewis Carroll's novel – and the surreal 'other' world that is describes – are the product of the author's ingestion of Victorian-era hallucinogenics. This theory, widely held amongst hippies of the late-1960s, was further popularised by the band Jefferson Airplane in the song *White Rabbit*, a US top-ten hit in the 'Summer of Love'. The opening verse runs:

> One pill makes you larger, And one pill makes you small,
> And the ones that mother gives you don't do anything at all,
> Go ask Alice when she's ten feet tall.

In an interview in November 2002 on the occasion of the song being voted 'Greatest Drug Song Ever' in *Mojo*, a British music magazine, Grace Slick, lead singer with Jefferson Airplane, commented:

> Parents were saying, 'Why are you taking all these drugs?' Well, they say the most important time in a child's life is between the ages of zero and five. Everybody reads their kids *Alice In Wonderland*, and there are drug references throughout it. Eat Me! Alice gets literally high, too big for the room. Drink me! The caterpillar is sitting on a psychedelic mushroom smoking opium. Peter Pan? You sprinkle some white dust – could that be cocaine? – on your head and you can fly! In every one of those stories you take some kind of chemical and have a great adventure. (Paytress: 2002)

An awareness of this non-literary background means the film's exhortation to 'follow the white rabbit' loses any sense of fairy-tale whimsy and assumes much darker overtones. The notion that the film intends these overtones is further substantiated in the film when Neo meets with Morpheus, leader of the rebels, and the person to whom the exhortation to follow the white rabbit ultimately leads. Morpheus (the keeper of 'dreams'?) offers Neo a choice. He outlines to Neo the story of Alice's fall down the rabbit hole, at the conclusion of which he offers Neo two pills, one red, one blue:

> Morpheus: 'This is your last chance. After this there is no turning back. You take the blue pill, the story ends, you wake up in your bed and believe whatever you want to believe. You take the red pill, you stay in wonderland, and I show you how deep the rabbit-hole goes.'
>
> Neo takes the red pill.
>
> Morpheus: 'Follow me.'

The exchange is clearly loaded with undertones of drug inducement, and indeed, minutes after taking the red pill Neo begins to observe the objects around him take on a sliding plasticity. From a filmic narrative perspective, this imagery serves the purpose of illustrating the beginning of Neo's grasp of the illusory nature of the Matrix. However, the evocation of hallucinogenic drug experience is clear.

Attuned to the drug-cultural subtext of the film, one begins to notice that in fact the entire film is laced with a dream-like, drug-induced visual quality, that some element of its appeal as a visual experience is in presenting the conscious mind with the mix of the soporific and the hyper-real that is characteristic of drug-experience. That the experience of immersion (it is the only adequate phrase) in the film's imagery is so pleasurable, and its enticement to drug-use so interwoven into that experience, creates another of the moral problems with the film. Neo's experience subsequent to taking the red pill is not only escapist, thrilling, dangerous, and exciting, but it ultimately confers invincibility on him. This, therefore, is not only an evocation of drug-usage without negative consequences, it additionally teeters into the territory of inducement, for it affirms the mendacious myths central to the allure of drug-culture.

III. Sexuality in *The Matrix*

The film's celebration of power and force, clearly evident in its idealisation of destruction, is also palpable in its fetishistic aesthetic. The main characters are all costumed in tight-fitting black clothing, black leather or glistening PVC. It is a clear evocation of sado-masochistic sexual subculture. The S/M subculture is driven by the eroticisation of dominance and by pleasure/pain interchange and exchange. *The Matrix* is suffused with such eroticised power and S/M sensibility. Feminist theologian Christine Gudorf's observations about the dynamics of S/M sex (Gudorf: 1994) provide insights which allow us to identify the sexual ideology running through *The Matrix*, and its coherence with the film's Nietzschean ethos.

Gudorf locates the eroticisation of dominance within our wider culture's construction of power, as expression of authority over, control over, demand from, and utility. As power is in our professional and social lives, so it is in our sexual lives.

> It should not surprise us that sex is distorted by understandings of power as dominance, because every other aspect of human life is similarly distorted. (Gudorf: 1994, 126)

The problem with the eroticisation of dominance, says Gudorf, is that it conditions us to experiencing sexual arousal in coercive sex, or sex-play which has the sense of coercion. The difficulty arising from such conditioning is that by knitting all the aspects of our lives into patterns of domination, we lose any critical distance from such patterns and thus the ability to identify what might be wrong with the exercise of dominative power (in all aspects of our lives) and with its consequences. In sexual terms, the eroticisation of domination becomes pointedly problematic if followed to the end-point, 'that force is the way to get a not-so-willing [person] turned on.' (Gudorf: 1994, 146)

Gudorf proposes that it is mutuality in sexual pleasure which should be normative – because justice demands it, intimacy depends on it, and pleasure requires it. However, she observes, it is the conditioning of our culture's eroticisation of dominance which remains actively normative for the majority of people,

> The problem is that there is often some very real conflict between what persons intellectually desire in sex – that which is both just and socially beneficial – and what actually turns them on. Unfortunately, the idea of sexual mutuality, and the image of mutual pleasuring, are not nearly so powerfully erotic for most people as are the idea and image of domination. (Gudorf: 1994, 124)

The interweaving of fetishistic clothing and an S/M aesthetic amidst its exultation of violence, clearly identifies *The Matrix* as pervaded with the eroticisation of dominance. The connections forged in our culture between pain, pleasure, and the exercise of power, are confirmed, affirmed and glamorised in *The Matrix*, and further re-presented to the viewer with an intensified alluring hyper-real gloss. Sex and dominative power are a natural conflation in *The Matrix*. Here, power, and the exercise of it, is sexy. Gudorf clearly identifies such a conflation, and the expression of sexual energies in dominative ways, as inimical to human (sexual) relational flourishing and fulfilment and a barrier to the possibility of genuine sexual intimacy. As a culture, then, we pay a heavy price for our indulgence of our prevailing sexual conditioning. *The Matrix*, in thrall to the exercise of power in all its forms, colludes with this widespread cultural sensibility, and by re-presenting it in such a heightened and thrilling manner, reiterates it as arousing, desirable and normative. By extension, the notions of mutuality, or of sexual egalitarianism, are thereby implicitly intimated to be erotically neutering.

IV. Philosophy and Spirituality in *The Matrix*

It is in regard to its theo-philosophical aspects that *The Matrix*'s flattery to deceive is most apparent. Much has been made in some quarters of the film's philosophical richness but, in truth, it would seem to simply offer a hi-tech re-working of some standard philosophical questions, particularly the 'brain-in-a-vat' hypothesis. This is certainly unusual in a populist entertainment, but would not seem to be philosophically meritorious in itself. Interestingly, much of the philosophical commentary on the film focuses on Cartesian and Platonic aspects, with little observation of the significance of the fact that the copy of Baudrillard's *Simulacra and Simulation,* in which Neo stores his illegal computer disks, deliberately flops open on-screen at the heading 'On Nihilism'. In the light of the foregoing analysis in this essay, this, and the Nietszchean sensibility with which the film is utterly permeated, would seem to offer more accurate insight into the concerns of the film than commentators' focus on Descartes and Plato.

Similarly, much has been made by some of the film's qualities as religious, specifically Christian, metaphor or fable. It is true that the consistent reference to Neo as 'the One', his own growing realisation of his 'true nature', and his resurrection by the character 'Trinity' would all seem to invite such a reading, and additionally, that Morpheus – who has been foretelling the coming of 'the One' – and Cypher – who betrays Neo – should be read as John the Baptist and Judas Iscariot respectively. Moreover, more detailed theologically-aspirational readings have also suggested that Neo's full name – Thomas 'Neo' Anderson – suggests a reference to 'doubting' Thomas, and an allusion to the title 'son of man'. In addition, the more isegetical have suggested that Neo's 'death', in the latter part of the film – lasting 72 seconds – should be interpreted as an allusion to the 72 hours that Jesus lay in the tomb.

In the light of the fact of these philosophical and theological speculations, it might be suggested that *The Matrix* is rightly admired, and should be lauded as a source rich with meaning and fertile with multiple readings. I suggest not. This is a film that through its complexity and allusion flatters to deceive, and by means of its deliberately ostentatious flirtation with the viewer's intelligence distracts from the awful coldness at its core. In the same way that the chapter heading in the Baudrillard text revealed the true philosophical stance of the film, despite the presence of other distractingly arrayed philosophical references, I contend that, despite the Christian allusions equally distractingly arrayed about, the spiritual outlook of the film is more truly captured in the following deliberate exchange between Neo and Choi, the character to whom he sells illegal computer disks:

Choi: 'Hallelujah, you're my saviour man. My own personal Jesus Christ.'
Neo: (warningly) 'You get caught using that...'
Choi: 'Yeh, I know. This never happened. You don't exist.'
Neo: 'Right.'

Perhaps the fact that the 'saviour' figure in this film trades in illegal computer disks, and asserts that what he needs to fulfil his mission are, 'Guns. Lots of guns', should be taken as an important indication that aspirations to finding any kind of genuine spiritual metaphor or meaning here are somewhat misguided, and that the film's flirtation with spirituality is entirely shallow and self-serving.

Mirror, Mirror...

The Matrix is a film which is fêted and admired in our culture on a number of levels, but its true expertise is in a kind of filmic sophistry, a sophistry that veils its disturbingly narcissistic nihilist core. It is in the light of such stark realisation that the viewer is 'unplugged' from the beguiling manipulativeness of the film, and *The Matrix* is revealed as not just a film about a tranquillising delusion which corrupts and disregards human values, but as itself a tranquillising delusion which corrupts, and disregards human values. That its duplicitousness has gone substantially unnoticed and unremarked, and that it continues to impact so powerfully on so many, internationally and cross-generationally, is profoundly unsettling.

In the mirror that is *The Matrix*, we are reflected as a culture saturated with violence, enthralled by the technologies of violence, and willing voyeurs of violence which seems to have no consequences, or at least no consequences for us. We revel in the pornography of violence. We have, additionally, eroticised dominance, displaced mutuality, jettisoned intimacy, and have ransomed ourselves to the lure and the rush of physical sensation – whether drug-induced or sexual – and with little regard for consequences. Unsurprisingly then, our engagement with spirituality is shallow and self-serving. What is reflected is neither pleasant nor reassuring, and I accept that as observations they are generalised and broadly drawn. However, I do contend that, as reflections of dominant aspects of our culture, they are neither inaccurate nor untrue.

Perhaps though, the insight that *The Matrix* most forcefully brings home is how much in thrall we are, as a culture, to appearances, to being deluded and distracted by the look of things, and to how things seem. That so much about *The Matrix* is cold, calculated, and cruel, yet the experience for so many of watching it is that it is intriguing, exciting, titillating, and captivating, and that it continues to be considered an appropriate source for sifting for insights into higher – sometimes transcendent – realities, is genuinely disturbing. As a culture, it would seem that if we are sufficiently visually stimulated and saturated, we will, consciously or unconsciously, set aside – surrender even – our engagement with those faculties by which we ordinarily, fundamentally as humans, make value judgements. We are, it would seem, slavish to visual stimulus.

These realisations are not novel, and it might reasonably be said that *The Matrix* considered as cultural mirror does not tell us anything about ourselves of which we are not already aware. This is true I think, to a degree, of those specifics which I have identified. However, I believe that what is being described herein is an instance where what is revealed in the sum is greater than what is revealed in the parts. That such a potent elevation of nihilism and alienation is so lauded in both popular and academic Euro-American culture is, I think, to reveal to us something significant of the shocking soullessness into which we have slid, and how dulled our higher sensibilities have become. That we are so easily seduced and sedated by the allure of appearance, to the willing neglect of ethical considerations, reflects starkly the ghastly impoverishment of our culture. *The Matrix* holds up a mirror to us, and in that mirror we can see the extent to which our ugliness has grown.

A Reflection Reflected On

It is neither a necessity nor an inevitability that technology and truly human values should be in opposition. The naturalist construction that sets up a polarisation between technology and nature / human nature is a false one. Our capacity for technology is neither alien to us, nor undermining of us but, like all our capacities, it is subject to choices of good implementation and expression. The situation is not even a matter of some technologies being good and others bad, since the laser beam can either annihilate cancer cells or guide a missile to annihilate the heart of a city, depending on the end to which it is directed. How we develop technology, implement it, and the place we assign in our culture to the products of it, are matters to be forged in the crucible of ethical enterprise.

The Matrix is a product of humans implementing technology and expressing themselves through the means of technology. It is an expression of human choice. The place it has been afforded in our culture is also an expression of human choice. In the future, we can make films that aspire to elevate the culture and us as humans in it (and these need not necessarily be 'serious') or we can make multiple versions and clones of *The Matrix, ad infinitum,* accept what that choice is telling us about ourselves and the culture we inhabit, and live with the consequences.

Bibliography

Baudrillard, Jean, (1994), *Simulacra and Simulation,* trans. Glaser, Sheila Faria, (Ann Arbor, MI: University of Michigan Press).

Baum, Frank L., (2000), *The Wizard of Oz,* (New York: Random House).

Carroll, Lewis, (2001), *Alice's Adventures in Wonderland,* (London: Bloomsbury).

Gibson, William, (1984), *Neuromancer,* (London: Voyager).

Gudorf, Christine, (1994), *Body, Sex, and Pleasure: Reconstructing Christian Sexual Ethics,* (Cleveland, OH: The Pilgrim Press).

Huxley, Aldous, (1994), *Brave New World,* (London: Flamingo).

Mallon, Matthew, (2003), 'Which Bit of the Future Would William Gibson Get So Terribly Wrong?', *Word,* Issue 5, July.

Orwell, George, (1998), *Nineteen Eighty-Four,* (London: Penguin Books Ltd.).

Paytress, Mark, (2002), 'High Priestess', *Guardian Unlimited,* Friday November 15. http://www.guardian.co.uk/arts/features/ story/ 0,11710, 840824,00.html

CHAPTER THREE

From Fear to the Beauty of Mystery

Cardinal Paul Poupard

Introduction

The basis for this article is the firm conviction that there is no inner contradiction between the viewpoint of any scientist and that of men and women who believe in God. Nor should there be any conflict between religion and science. Those who say there is are rehearsing a tired and unconvincing litany of inaccuracies. Suffice it to say that the present Pope made it clear from early on in his pontificate that, '... the church freely recognises ... that it has benefited from science' (John Paul II: 1979); on several occasions he has drawn attention to the fact that 'collaboration between religion and modern science is to the advantage of both, and in no way violates the autonomy of either' (John Paul II: 1979).

This conviction is at the heart of the Encyclical Letter *Fides et Ratio*, and it is worth remembering that the Pope told everyone who took part in the Jubilee of Men and Women from the World of Learning, in May 2000, that 'the church is not afraid of science'. This is a conviction I wholeheartedly share, particularly since the days when the Holy Father asked me to work on the commission set up to re-evaluate the Galileo case, 'with the aim of improving relationships between the church and science' (Béné: 1987, 177).

The Second Vatican Council

The fathers of the Second Vatican Council recognised the real challenge posed to a culture's customs and conventions by technological advances. When a culture begins to feel the influence of sophisticated developments in science and technology, questions of communal and individual identity and continuity make it important to find a way of harmonising these new elements with long-established ones. 'How is the dynamism and expansion of a new culture to be fostered without losing a living fidelity to the heritage of tradition', they asked (*Gaudium et Spes:* 1996, 56).

There is a broad sweep of reactions to technological innovation, ranging from worry and fear to confident optimism: I should like to take a brief look at the two extremes, and suggest two beacons by means of which it is possible to take one's bearings in this area.

Fear of Technology

The English language has a marvellously evocative phrase that alerts the hearer to a diffidence on the part of the speaker towards anything other than what is tried and trusted. There is no mistaking the stance of the person who describes something as 'new-fangled'. It could be disdain, but I suspect it often masks bewilderment or fear – fear of the unknown, or fear of what one is unable to understand or control. It is an attitude one expects in older generations, but it is not exclusive to them.

There is one approach that makes a fine distinction between technical innovation and technology, pointing to an almost imperceptible semantic shift over time. There are examples of technical genius from every age, in the inventions that enable men and women to perform the tasks they need to accomplish in order to make life easier, more comfortable or more enjoyable. But now there is 'high technology', 'state-of-the-art' technology and the like. Some detect the replacement of technical genius by an ideology of technique – techno-logy. Technique as such poses no threat, whereas there are dangers when technology is imposed at all costs, as if it gave an answer to every question, in a kind of utopian vision. This is the conviction behind Aldous Huxley's use, at the beginning of his novel *Brave New World,* of the following quotation from Nicholas Berdyaev:

> Utopias are possible. Life is marching towards utopias. Perhaps a new age is beginning, an age where intellectuals and the cultured class will dream of ways of avoiding utopias and returning to a non-utopian society which is less perfect and more free. (Huxley: 1994 (a), Frontispiece)

Brave New World offered a chilling view of how things might be in a technological utopia. The historical background is significant: the Soviet empire was very firmly established, and the Nazis were emerging from the wings elsewhere. What Huxley wrote was not really a *utopia,* but much more a *dystopia,* the analysis of a humanly dysfunctional world when inhuman or monstrous forces have been let loose.

This is exactly the root of the terror engendered by Mary Shelley's novel *Frankenstein,* to which she referred as her 'hideous progeny' in her introduction to the second edition, after thirteen years of reactions to the book's first appearance. The heaviness of the book lies in the real location of evil. The monster created by Frankenstein is fierce and extremely destructive. But what is really problematic is Victor Frankenstein's own way of thinking. 'I succeeded in discovering the cause of generation and life', he said; 'I became myself capable of bestowing animation upon lifeless matter' (Shelley: 1993). This is very much the stuff of adventure stories of those days, what we might now call science fiction. But he went further,

> No one can conceive the variety of feelings which bore me onwards ... A new species would bless me as its creator and source; many happy and excellent natures would owe their being to me. No father could claim the gratitude of his child as completely as I should deserve theirs (Shelley: 1993, 43).

Technological inventiveness always has a context, and Mary Shelley points to the frightening consequences of self-centred research and development. There may be something of the Luddite in her, as in all of us, but she allows us to see that the question 'why?' is very important. In this case, Victor certainly sought power and gratification. The novel also raises a question that arises in many other contexts, like the safety of nuclear power. To what extent can technology run out of the control of its inventors or operators?

Something much closer to home, for many in richer, developed nations is the control of the Internet, particularly in the case of children and those who are morally very weak. Who is really in charge? What is the Internet actually doing to human relationships? All sorts of theories have been put forward on this, but it is probably too soon for us to know with certainty how to answer those questions. I have been fascinated to hear what experts have to say on the large numbers of people involved in the creation of multiple virtual selves in cyberspace. Who do people think they are? I am not surprised that there are fearful reactions to all this.

A Positive View of Technology

Earlier, I mentioned Huxley's use of Berdyaev's idea that a 'new age' might be dawning, when the terrifying prospect of technologically perfect but inhuman utopias might at last recede. The same positive note was sounded, from a completely different approach, in the Second Vatican Council. There it is said that a new age of human history has come into being as a result of profound social and cultural changes in the circumstances of the lives of people today.

> New ways are open, therefore, for the perfection and the further extension of culture. These ways have been prepared by the enormous growth of natural, human and social sciences, by technical progress, and advances in developing and organising means whereby men can communicate with one another' (*Gaudium et Spes:* 1996, 54).

Clearly, the Council fathers wanted to give a very upbeat, positive evaluation of the contribution of the sciences to the life of the human race.

At the very least, I would like to say, with Heidegger, that:

> ...what is dangerous is not technology. Technology is not demonic; but its essence is mysterious ... The threat to man does not come in the first instance from the potentially lethal machines and apparatus of technology (Heidegger: 1978, 309).

For Heidegger, the problem with technology is much more a question of finding ourselves prevented (he uses the technical term *enframing*) from considering some more basic truth. He paraphrases Hölderlin's poem *Patmos*: 'We look into the danger and see the growth of the saving power' (Heidegger: 1978, 315). In simpler terms, he admits that technology can and does fascinate or even entrance us. But, as long as we realise that that is happening, we also still know that there are more fundamental questions to ask. I am convinced Heidegger is saying in dense poetic language what Jesus said about wise patience in the parable about the tares and the wheat in chapter 13 of Saint Matthew's gospel (Mt 13:24-30). He is right, since it is naturally very difficult for us to see how to disentangle the positive and negative elements of technology.

The Role of Conscience

In 1979, Pope John Paul II said that 'the search for truth is the fundamental task of science' (John Paul II: 1979) and he insisted that scientists should never be deprived for political or economic reasons of their freedom in the area of research. Science applied practically is fully developed in the sphere of technology.

> Applied science should be allied with conscience, so that, in the triad, science-technology-conscience, it may be the cause of the true good of humankind, whom it should serve (Béné: 1987, 195f).

The Pope reinforced this point in the Encyclical *Fides et Ratio*, when he urged scientists to:

> ... continue their efforts without ever abandoning the sapiential horizon within which scientific and technological achievements are wedded to the philosophical and ethical values which are the distinctive and indelible mark of the human person (John Paul II: 1998, 106).

Again, more recently, in his Apostolic Letter marking the end of the Great Jubilee of the year two thousand, he addressed these words to scientists:

> Those using the latest advances of science, especially in the field of biotechnology, must never disregard fundamental ethical requirements by invoking a questionable solidarity which eventually leads to discrimination between one life and another and ignoring the dignity which belongs to every human being (John Paul II: 2001, 51).

It is interesting that what the Holy Father says about the essential role of conscience in the work of scientists, and particularly in the sphere of technology, does not apply solely to people of strong religious convictions. The reason he insists on conscience being given its place is that it allows science and technology to respect the dignity of the human person. This is an appeal for what, in other words, would be the building blocks of a thoroughly human culture, a call for even the most highly qualified researcher to accept that his or her work has limits, principally because it has always to serve the common good of humanity.

The Beauty of Mystery

The present Pope sets great store by culture. Indeed, it was he who set up the Pontifical Council for Culture in 1982. His vision of culture is a broad one, and part of the remit of the Council is to foster the dialogue between science and faith. This fits perfectly into the vision of the Second Vatican Council, where the human person is seen to be capable of enormous creative effort, and society is enriched when men and women of varied intellectual capacities and backgrounds pool their efforts. When this happens, a man or a woman 'can do very much to elevate the human family to a more sublime understanding of truth, goodness and beauty' (*Gaudium et Spes:* 1996, 57). The wisdom and relevance of this vision was brought to mind at the recent Consistory of Cardinals in Rome, when Cardinal Godfried Danneels of Malines-Bruxelles said that people today are 'hesitant before the true, resistant to the good, but captivated by the beautiful'.

Towards the end of his essay on technology, Heidegger says this:

> There was a time when it was not technology alone that bore the name *techne*.... Once there was a time when the bringing forth of the true into the beautiful was called *techne*. The *poiesis* of the fine arts was also called *techne*' (Heidegger: 1978, 315).

Those days are long gone; they were the days when knowledge, later called *scientia,* had not yet split into the myriad specialisations there are in universities and schools today. Perhaps the challenge in Heidegger's vision of technology is to see the link of the scientific craft to the creative forces of art, and thus to understand that there is a certain beauty in the capacity that drives technology forward.

To see a link between the artistically poetic and the technologically poetic opens the mind and the heart to a different vision of the world. For me, as a Christian, it speaks essentially of being involved in a prolongation or an extension of God's creative work. That speaks to me of my limits, my obligations, but also of the heights to which God's grace can raise my creative powers. When scientific research is conceived as totally independent in the sense of having no link with God, when there is no reference to the Creator,

> Anyone who acknowledges God will see just how false such a meaning is. For without the Creator the creature would disappear ... When God is forgotten ... the creature itself grows unintelligible' (*Gaudium et Spes:* 1996, 36).

Those words, taken from one of the Second Vatican Council's main documents, are in perfect harmony with Mary Shelley's reflection in her introduction to *Frankenstein*: 'supremely frightful would be the effect of any human endeavour to mock the stupendous mechanism of the creator of the world' (Shelley: 1993, 4).

In *The Doors of Perception,* Aldous Huxley explores the experience of altered states of consciousness brought about by the use of psychedelic drugs. Towards the end of the book, he includes the following uncharacteristic reflection,

> Near the end of his life Aquinas experienced Infused Contemplation ... Compared with *this*, everything he had read and argued about and written ... was no better than chaff or straw ... For Angels of a lower order and with better prospects of longevity, there must be a return to the straw. But the man who comes back through the Door in the Wall will never be quite the same as the man who went out. He will be wiser but less cocksure, happier but less self-satisfied, humbler in acknowledging his ignorance yet better equipped to understand the relationship of words to things, of systematic reasoning to the unfathomable Mystery which it tries, forever vainly, to comprehend' (Huxley: 1994 (b), 55f).

Creativity allied to an awareness of one's own creatureliness opens our mind to the acknowledgement of the Mystery of which Huxley wrote. It is the more profound mystery hinted at in the mysterious element of technology. It can inspire fear, but should really inspire nothing more than humility, which I believe is best translated in English as 'being down to earth'. If even the most highly qualified researchers are humble in that sense, the fruits of their work should hold out no threat to their fellow men and women. On the contrary, they should be an enrichment to them as individuals, and a cement for a more human culture. This brings me to think yet again of the urgent need for a constant and courteous dialogue between people of faith and those involved in scientific research. Permit me to conclude simply by quoting a short excerpt from the Second Vatican Council's Pastoral Constitution on the Church in the Modern World. It is not an answer to all the questions we may have, but I think it is a very encouraging piece of advice.

> May the faithful live in very close union with the other men of their time and may they strive to understand perfectly their way of thinking and judging, as expressed in their culture. Let them blend new sciences and theories and the understanding of the most recent discoveries with Christian morality and the teaching of Christian doctrine, so that their religious culture and morality may keep pace with scientific knowledge and with the constantly progressing technology. Thus they will be able to interpret and evaluate all things in a truly Christian spirit' (*Gaudium et Spes:* 1996, 62).

Bibliography

Béné, Georges (1987), 'Galileo and Contemporary Scientists' in, Poupard, Paul, (ed.), *Galileo Galilei: Toward a Resolution of 350 Years of Debate – 1633-1983*, (Pittsburgh: Duquesne University Press).

Heidegger, Martin (1978), 'The Question Concerning Technology' in, Krell, David Farrell, (ed.), *Basic Writings: Martin Heidegger*, (London: Routledge).

Huxley, Aldous (1994(a)), *Brave New World*, (London: Flamingo).

Huxley, Aldous (1994(b)), *The Doors of Perception*, (London: Flamingo).

John Paul II (1979), Address to the Pontifical Academy of Sciences at the Commemoration of Albert Einstein.

John Paul II (1998), *Fides et Ratio*.

John Paul II (2001), *Novo Millennio Ineunte*.

Shelley, Mary (1993), *Frankenstein*, (Ware: Wordsworth Editions Ltd.).

Vatican Council II (1996), 'Gaudium et Spes', in, Flannery, Austin (ed.), *Vatican Council II: The Conciliar and Post-Conciliar Documents*, Vol. I, (Dublin: Dominican Publications).

CHAPTER FOUR

Mind The Gap: The Tarnishing of a Transcendent Technology

Paul Brian Campbell

Introduction

In 1999, Michael Breen and I co-wrote a paper, 'The Net, Its Gatekeepers, Their Bait & Its Victims – Ethical issues relating to the Internet,'[1] in which we compared the promise of what was then still described as a 'new medium' to the reality of the media oligopoly-driven marketing tool that it rapidly appeared to be becoming.[2] It is time to revisit the dilemmas we identified in 1999 and to highlight new concerns that have since arisen. Perhaps the biggest change that has occurred in the last four years is the change in the monetary value of the 'Internet economy.' Nathan Newman points out that the Internet became, 'the defining economic event of the end of the twentieth century – a fact reflected by the obsessive media attention and by the raw economic explosion of companies associated with it' (Burnett & Marshall: 2003, 61). As we all know, that 'economic explosion' ended in the severe downturn in the stock market which began in April 2000, but the Internet itself continues to increase in size and ubiquity.

The Internet has succeeded in removing 'the geographical and time limitations of operating in a global economy' (Burn & Loch: 2002), allowing people to communicate in new and unforeseen ways. M. Godwin reminds us of the unique quality of the Internet:

> The Net is the first medium that combines all the powers to reach a large audience that you see in broadcasting and newspapers with all the intimacy and multi-directional flow of information that you see in telephone calls. It is both intimate and powerful (Burnett & Marshall: 2003, 61).

Although we do well to heed warnings that we are too close to unfolding developments to be able to make useful judgments,[3] we cannot simply ignore the remarkable eruption of the Internet and we are certainly better able to assess the situation than we were four years ago.

We are still surrounded by both utopian and apocalyptic visions arising from the headlong rush to go online, but there seems to be a growing awareness that the reality is to be found somewhere in the middle – there are both real costs as well as genuine benefits stemming from the Internet. Although its optimistic boosters may take pride in such developments as the first Swiss online election in March of 2003, critics of the Internet point out that, 'exploitation, exclusion, eco-vandalism and authoritarianism continue to pervade the planet: enabled, in most cases, by the technological infrastructure of the information society' (Hornby & Clark: 2003, 43). John Cassidy, author of *dot.con: the greatest story ever sold,* has also pointed out how world events have helped to refocus our perceptions:

> The promise of the Internet wasn't just technological: it was also ideological. Once digital networks had liberated them from the confines of tradition and physical location, human beings would come together and transcend ancient divisions: tribal, religious, racial, and economic. After September 11, it seems ludicrous to speculate about an escape from history or geography (Cassidy: 2002, 313).

Several commentators are also becoming more sanguine about the prospects of an immediate upheaval in existing power structures resulting from our adoption of the Internet. W. Edward Steinmueller states that, 'The virtual economy does not provide a fundamentally new approach to the issues of procedural and institutional authority, although it provides many new tools for exercising and articulating these types of authority' (Mansell: 2002, 53). Robert Burnett and P. David Marshall agree with this analysis and note that the Internet's 'structure makes the centre harder to determine; yet there are clear directions in the flow of information that both replicate and represent existing power structures that are part of both globalisation and what has been labelled as the network society' (Burnett & Marshall: 2003, 44). Cassidy notes that, 'It is difficult to think of a single example in which an internet company has successfully supplanted a major old economy firm' (Cassidy: 2002, 316). He also cites the views of Robert Gordon, an economist at Northwestern University, who places the impact of the Internet within an historical perspective:

> Internet surfing may be fun, but it represents a far smaller increment in the standard of living than achieved by the ex-

tension of day into night by electric light, the revolution in factory efficiency achieved by the electric motor, the flexibility and freedom achieved by the automobile, the saving of time and the shrinking of the globe achieved by the airplane, the new materials achieved by the chemical industry, the first sense of live two-way communication achieved by the telephone, the arrival of live news and entertainment into the family parlor achieved by radio and then television, and the enormous improvements in life expectancy, health and comfort achieved by urban sanitation and indoor plumbing.

The Internet is a revolutionary means of communication, but it hasn't made people live longer, changed where they live, or made it any easier for them to get from Paris to New York (Cassidy: 2002, 320).

The Dimensions of the Internet

When Breen and I wrote in 1999, the Internet was expanding at a mind-boggling pace. Recent evidence suggests that the massive growth spurt, when usage percentages doubled at least every twelve months, has begun to slow down – at least in parts of the developed world. NUA, an Irish-based major source for Internet trends and statistics, says that its best guess, as of September 2002, is that 605.6 million users worldwide were on the Internet.[4] Using its data, as we did in 1999, allows us to compare Internet usage across the world:

Table 1: Million Internet Users by Continent: 1996, 1999 & 2002[5]

	1996	1999 [March]	2002 [September]
Africa	-	1.14	6.31
Asia/Pacific	6	26.55	187.24
Europe	9	36.11	190.91
Middle East	-	0.78	5.12
US/Canada	30	94.2	182.67
Latin America	-	4.5	33.35

The five countries with most citizens using the Internet are: the United States, Canada, South Korea, the United Kingdom and Japan. (Marino, *The New York Times*, 16 February 2003, 3:9:1)

In Japan, largely fuelled by increases in broadband connectivity and access to the Internet via mobile phones, between 2001 and 2002, the percentage of the online population grew more than 10% to reach 67% (*Agence France Press,* 20 November 2002). South Korea, with 53% online in 2002 saw an 8% growth rate over two years (*Agence France Press,* 20 November 2002). In Britain, the online percentage remained at approximately 42% throughout 2002[6] and Canada's 62% online population, saw barely any growth (*Online Reporter,* 12 December 2002). In the United States, the country with the most people online, the UCLA Internet Project reports that the percentage of people who accessed the Internet in grew to 72.3% in 2001, but fell to 71.1% in 2002.[7]

It remains to be seen if the dip in Internet use in the United States is a temporary phenomenon or if it will be replicated in any of the other 'top five' countries. Growth in Internet access remains strong in developing nations but, as Table 1 indicates, many countries have a long way to go before internet access becomes readily available to a sizeable percentage of their populations. As Louis Gerstner, the former CEO of IBM, points out, 'the fact remains that more than half the world's population have yet to make a phone call. The half-billion Internet users … [are] impressive for a technology that's still in its infancy, yet represents less than 10 percent of the people on the planet' (Gerstner: 2002, 342).

The Digital Divide

Pippa Norris has usefully identified three differing aspects of the digital divide: a *global divide* between industrialised and developing countries, a *social divide* between the 'information rich' and 'information poor' within nations, and a *democratic divide* which separates those who proactively use the Internet for civic and political purposes as opposed to those who passively ingest its contents.[8]

In our 1999 article, Breen and I had a section titled 'Africa could use a good net …' It would seem that it is still waiting for the Internet to develop into a major medium of information exchange.

It is startling that the highest figure for the whole continent doesn't even reach 12% of the population. Most African countries, in fact, including Egypt and Nigeria, have less than 1% of

Table 2: Current Top Five Internet Users in Africa [9]

Country	# of users	% Population
Seychelles	9,000	11.24
South Africa	3,068,000	7.03
S. Tome & Principe	9,000	5.28
Tunisia	4,000	4.08
Cape Verde	12,000	2.94

their citizens with access to the Internet. Other developing countries report similarly low numbers: Brazil has 7.8% of its population online, Russia is at 5.1%, China has 3.58%, Mexico is at 3.38%, India is at 0.67, Bangladesh has less than 0.2%, Iraq is at 0.05% and some of the world's poorest nations, such as Afghanistan and East Timor, do not have any registered percentages at all.[10]

These fairly dismal numbers seriously challenge the notion that the Internet can act as a great force for democratisation and that barriers to economic advancement and social development are swept away by the free flow of information available on the Internet. It should be noted that, even should access to the Internet suddenly become freely available across the world, illiteracy would still be a major stumbling block. Further, English is still very much the Internet's dominant language, but is spoken by less than 10% of the world's population. The English-speaking world still has more than 80% of the top-level Internet sites and hosts close to 80% of Internet traffic.[11]

Although the United Nations General Assembly declared in June 2002 that, 'The digital divide threatens to further marginalise the economies and peoples of many developing countries,' this opinion is not universally accepted. The quotation from the General Assembly is cited by Charles Kenny in an article entitled: 'Development's False Divide: Giving Internet Access to the World's Poorest Will Cost a Lot and Accomplish Little' (Kenny: 2003, 76). Certainly, the steps being taken by the developed world to help break down the 'global divide' seem fairly timid. In March of 2003, for example, the Bush Administration announced a 'digital divide initiative' costing $2 million over three years to send volunteers from major telecommunications companies to Senegal to help 'turn existing networks – such as cyber cafes – into better resources for entrepreneurs and small businesses ... If successful, the program could be rolled out to another 20 countries over five years, officials said' (*Associated Press Online*, 3 April 2003. LexisNexis).

The social divide between the 'information rich' and the 'information poor' is apparent in every country with access to the Internet. This part of the digital divide has been, and remains, of great concern in the United States even though it is the country with the highest percentage of its citizens with access to the Internet. With the falling costs of computers, more people have installed them in their homes. The Corporation for Public Broadcasting released a study in March 2003 which found that:

> The 'digital divide' between rich and poor children in the US is shrinking as youngsters of varying income levels and races are increasingly using the Internet ... However gaps still exist as white children and those from rich families are still more likely to have high speed Internet access at home ... More than two-thirds of low income households now have a computer at home compared to 98 percent of high income households ... Two years ago, only half of low income households had a computer while 9 out of 10 rich households did ('Communications Today,' 3 March 2003, v.9 i50. Lexis Nexis).

Despite this good news, however, it should be noted that one third of all low income households (which tend to be in rural areas or among central city minorities), are still without computer access at home and have to rely on free access at schools or in libraries. The playing field may not be quite so dramatically pitched as before, but it is by no means even.

Norris' notion of the democratic divide is a fairly novel one. It is clear, nonetheless, that a limited number of people proactively use the Internet for civic and political activities. As with television viewing habits, the vast majority of the population seem to be content with what is presented to them through the most popular channels of distribution.

If the results of the UCLA Internet Project hold for Internet users across the world, the Internet may not be transforming society as much as its advocates might wish. In 2002, most of those surveyed reported that the amount of time and the kinds of interactions they and their children had with family and friends had changed little (except for television in a group). Further, only about half of the respondents agreed or strongly agreed that the Internet had helped them increase the number of people with whom they were in contact.[12]

Internet Traffic

In 1999, when Breen and I listed the Top Ten Internet Service Providers [ISPs] they were all based in the United States. Their location hasn't moved, but the order of popularity has changed somewhat:

Table 3: The Top Ten Internet Service Providers / Web Portals

Rank	1999 Site	2003 Site
1	Yahoo	Yahoo
2	AOL	Google
3	Microsoft	MSN.com
4	Netscape.com	AOL.com
5	Geocities.com	AskJeeves.com
6	Excite	Overture
7	Lycos	Infospace
8	MSN	Netscape
9	Infoseek	Altavista
10	Altavista	Lycos

Yahoo is still at the top of the list but its search capabilities, along with those of AOL, are now provided by Google, a privately-held company which didn't exist in 1999. Netscape, formerly the fourth-rated site but bruised from its rivalry with Microsoft, has been knocked off the list, as have Excite and Infoseek. Overture, now in sixth position, has just purchased ninth position Altavista, and Yahoo, which has already bought the Inktomi search engine, recently issued a $750 million bond offering so that it can make further acquisitions. As in other areas of the communications industry, consolidation seems to be the order of the day. Gord Hotchkiss sums up the situation in this way:

> The industry is maturing and the major players marking [sic] their territory in anticipation of an all out turf war. Revenue

> models are starting to work and search engines are beginning to make money. Pay per click and paid inclusion have turned around bottom lines for the biggest players, and now that positive revenue appears possible, the competitive market is starting to consolidate. Make no mistake about it. Yahoo, Google, Overture and MSN are betting that search will turn into a multi billion dollar industry (currently, total search engine marketing revenues are about 1.5 billion). What's happening now is analogous to a chess match entering the end game. The pawns have been sacrificed and the power pieces are now engaged in a complicated game of attack and counter attack.[13]

The end result for the Internet user is likely to be a diminution of choice and a more confined, advertising-driven online environment.[14]

Already Internet surfers have to deal with far more than advertising banners, 'pop up' and 'pop under' advertising pages and animated commercial messages which invade part or all of the screen. 'Hidden sponsorships' on Yahoo, MSN and other sites leave it unclear that the first page of some results from searches are from companies who have arranged for either 'paid inclusion' or 'paid placement.' Site sponsoring, where a company pays for 'tenancy' of an entire site, can be even more insidious because on 'journalistically oriented web pages, the line between editorial content and advertisement is blurred' (Seebacher: 2002, 24). This is particularly worrying because, according to the UCLA Internet Project, 'The more than 70 percent of Americans who use the Internet now consider online technology to be their most important source of information, ranking the Internet higher as an information source than all other media including television and newspapers…' (*Ascribe Newswire,* 30 March 2003. LexisNexis).

Web portals and search engines are not the only part of the Internet that have been inundated by commercialism. One of the most remarkable innovations brought about by the Internet was the ability to exchange e-mail. It did not take commercial interests long to discover that e-mail could be used as a marketing tool and now our inboxes are often stuffed with unsolicited commercial e-mail [UCE], more commonly known as 'spam'.

US Senator John McCain, at a May 2003 Hearing of the Senate Commerce, Science and Transportation Committee, has succinctly summed up the current situation:

> Less than two years ago spam made up only eight percent of all e-mail. Today industry experts estimate that more than 45 percent of all global e-mail traffic is spam and many expect it to reach the 50 percent mark by this summer. AOL estimates that it blocks 80 percent of all its inbound e-mail – nearly 2.4 billion messages each day. Managing this influx adds real cost to consumers and businesses. There are other costs to Americans such as the cost to our children who may be victimised by the nearly 20 percent of spam that contains pornographic material some including graphic sexual images. The FTC also tells that two-thirds of all spam contains deceptive information, much of it peddling get rich quick schemes, dubious financial or health care offers and questionable products and services (*Federal News Service,* 23 May 2003. LexisNexis).

At the same hearing, it was pointed out that the UCE problem has grown so severe that it threatens to engulf and disrupt the entire e-mail system. Given this danger, it is all the more remarkable to learn from Senator Chuck Schumer that 90% of spam originates from just 250 Internet users (*Federal News Service,* 23 May 2003. LexisNexis). The technology of the Internet can lead to real transcendence, but it is also extremely vulnerable to attack.

The Internet has evolved a long way from its roots as a file sharing medium for the military and the academic community. A recent survey of 130 US companies indicates consumers spent $76 billion shopping online in 2002. This is a 48 percent increase over 2001, and the projections for 2003 are estimated at $100 billion, which represents 4.5 percent of total retail sales. Consumers, however, are often handing over to the retailers a lot more than just their money.

Privacy Concerns

There is nothing new about online privacy concerns. Internet privacy has been defined as 'the seclusion and freedom from unauthorised intrusion' (Dhillon 2002, 67). In 1999, Breen and I wrote about 'cookies' – those files placed on computer hard drives that allow for the tracking of Internet surfing. Savvy Internet users, however, have always been able to set their Internet browser preferences to warn them when cookies were being requested or they could arrange for some or all cookies to be blocked. Increasingly, however, websites are refusing access unless the cookies function has been activated and, despite the oft-cited ability of Internet users to be anonymous, it has become virtually impossible for Internet users to remain unidentified.

A few particularly egregious attacks on privacy have been turned back. Real Networks, a widely-used media player, was forced to apologise for the application built into its RealJukeBox which allowed it to capture information about a consumer's musical preferences. In 2000, DoubleClick, one of the Internet's most successful advertising companies, came under fire when it bought a mail-order purchase database and attempted to match it with its own vast consumer database (Dhillon 2002, 2).

There has, however, been a steady increase in the installation of 'adware' and 'spyware' onto personal computers as a result of software downloads. When consumers install programmes that allow them to swap media files, organise their desktops, help them remember their online user names and passwords or even transform their cursors into animated characters, they most often automatically choose 'I agree' on the software licensing terms. Had they spent time reading the dense legalese of the terms and conditions of accepting the software, they would have discovered that they'd voluntarily surrendered a great deal of personal privacy.

Both 'adware' and 'spyware' collect information about users; 'adware' generally makes its presence felt with 'pop-up' and other kinds of advertising but, as its name suggests, 'spyware' secretly gathers information and sends it out to advertisers and other data harvesters without making its intent apparent. Bonzi Buddy, for instance, is a programme which installs a friendly interactive purple ape on your desktop; it not only reports on your Internet activity, but will constantly be urging you to purchase something from an online vendor. Other downloads, such as those 'bundled with peer to peer management programmes like Kazaa have actually hijacked user computer's unused processing power and siphoned it off to be used in a virtual networked mega-processor. And, in the worst case, downloaded software can actually act like a virus or Trojan horse and install malicious software, or 'malware', onto your computer.'[15] If people were more aware of how compromised their online privacy has become, perhaps they would not show quite such enthusiasm for online pornography and other personally revealing web-surfing activities.

Pornography

In 1999, Breen and I reported that pornography was rife on the Internet. We entered the terms 'pornography' and 'porn' into the Dogpile search engine and they yielded a combined total of more than 790,000 hits. Replicating the same search for this study, I was immediately directed to a pornographic site where a credit card number was required in order to proceed. Entering the terms instead into Google resulted in a combined total of more than 69 million hits. This should come as no surprise, given that the worldwide pornography business was estimated in 2001 as being worth $56 billion per year (Burnett & Marshall: 2003, 114).

According to Alexa Research, '1 in every 300 searches was for 'sex' and other terms such as 'porn,' 'nude,' 'XXX,' and 'erotic' were all among the highest twenty' searches on the Internet (Burnett & Marshall: 2003, 114). Mark Kastleman, author of *Drug of the New Millennium*, claims that 'Ninety-five million Americans a month are looking at pornography on the Internet' (Cartwright, 'The Utah Statesman' 9 March 2002 LexisNexis). If all of these visitors were adults, one could legitimately debate whether or not we are discussing an ethical issue. Studies, however, appear to show that a considerable percentage of those visiting such sites are children. According to Nielsen/Net Ratings, however, nearly 16 percent of visitors to adult-oriented sites in February 2002 were under the age of 18.[16] Meanwhile, the Australia Institute announced that, '84 per cent of teenage boys spent time watching material on pornographic sites ...' ('The Advertiser,' 6 March 2003. LexisNexis).

Despite moves in several US states to install filtering software in libraries and other places where the Internet is publicly accessed, it will be impossible to prevent children finding pornographic materials on the Internet. Ironically, perhaps, it is the increasing commercialisation of the Internet that may best protect children because they will be unable to pay to view pornographic materials.

In 2002, there was a great deal of media coverage on child pornography, generated in particular by the arrest of a number of high profile figures. This followed a number of simultaneous police raids across the world, including Operation Candyman in the US, Operation Wonderland in Britain, and Operation Amethyst in Ireland. These operations followed the successful closure of a major child pornography internet site in the US. Customers of the site were traced via their credit card numbers. The discovery that the US has provided the UK and Ireland with evidence against those who purchased abusive images emphasised for many that the Internet is not as anonymous as many thought it was and may well cause people to think twice before using their credit cards for such purposes, thus diminishing the revenues that can be made from such illegal activities.

Gambling

Another 'vice' that has established itself online is gambling. By the close of 2001, approximately 52 million people worldwide were engaged in Internet gambling and, according to Bear Stearns, the revenues for 2002 were expected to exceed more than $8 billion. Perhaps the most staggering figure of all, however, was that online casinos were set to keep about 75% of the money deposited (*M2 Presswire,* 10 February 2003. LexisNexis). Consumers, obviously, are not expected to do well in this equation.

There are nearly 12 million people in the United States who bet online, despite the fact that such activity has been declared illegal by the US (Shinkle, 'St Louis Post-Dispatch, 4 May 2003, A1. LexisNexis). Despite the banning of Internet gambling-related credit card transactions by US banks, gamblers are still sending money directly from their bank accounts or are wiring money to their offshore bookies.

Online gambling remains legal in the European Union and has become increasingly popular: in June 2002, visitors to online gambling sites increased 15% in the UK, by 18-19% in Italy and Germany, by 25% in Spain and by a whopping 31% in Norway. As with access to pornography, however, a significant amount of Internet gambling traffic is attributed to youngsters. It has been estimated that 16.8% of those June 2002 visitors were aged seventeen and under.[17] These figures are especially important given that a study in 'Psychology of Addictive Behaviors' reveals that 'more than 15% could be described as having a pathological problem.'[18]

Internet Crime

In the early days of the Internet, there was relatively little criminal activity and what did occur centred around hackers, virus writers and those who distributed and used child pornography.

All those crimes have grown in scope and number and have been joined by increasing incidents of computer sabotage, data and identity theft, attacks on networks and ideologically extreme websites.

In 2001 the Confederation of British Industry announced that two thirds of UK companies had suffered 'a serious security incident, such as hacking, virus attacks, or credit card fraud in the past year.'[19] In 2002 'Newsbytes', an online news organisation, reported that almost 90% of US businesses and governmental agencies had suffered hacker attacks.[20] The situation was so grave that later in 2002, the German Interior Minister was led to state that 'Internet Crime has developed into a clear and present danger for information-based societies' (*Agence France Press*, 22 December 2002. LexisNexis). Noting that more that 48,000 Internet-related fraud complaints were made in 2002, a US Department of Justice press release stated that, 'Internet Fraud and Abuse is one of the most rampant forms of white collar crime' (*FDCH Federal & Agency Documents*, 16 May 2003. LexisNexis). If CNN is to be believed, however, the situation has gone from bad to worse. It reported on 16 May 2003 that 'Since the beginning of the year, the government has conducted investigations involving more that 89,000 victims who lost a total of $176 million.'[21]

It has been plausibly suggested that the increase in Internet crime is a result of the recent economic downturn. It is also relatively easy for those who are technically skilled to steal money and data from their employers. It is estimated that 70% of commercial computer intrusions are caused by employees, but that the chances of catching and successfully prosecuting Internet criminals are fairly slim because of the difficulties of proving guilt without fungible evidence (Tedeschi, 'The New York Times,' 29 January 2003, p. 1 Lexis Nexis).

Conclusion

What are we to make of the state of the Internet in 2003? The initial hype about how it would break down barriers and allow peace and democracy to flourish across the world now seems more than a little extravagant, as do the claims that it would knock down hierarchies and clear out oligarchies by presenting the same information to everyone. As we have seen, however, the Internet currently tends to enhance established power structures and is very much oriented in favour of those who advocate globalisation. It was, after all, Tim Berners-Lee, the man who conceptualised the World Wide Web, who said, 'I told people that the Web was like a market economy' (Cassidy: 2002, 21).[22]

Market economies reward a few very handsomely, work in a decent fashion for some and keep others at a distinct disadvantage. The Internet increasingly presents its users with the same attractions, opportunities, temptations and dangers as anywhere in the 'real world'. People are using the Internet both for the most transcendent purposes, but also for the most tarnished.

Endnotes

1. Paper for the 'Media and the Marketplace: Ethical Issues' Conference at Mater Dei College, Dublin, Ireland, in February 1999. The paper was later included in Cassidy, E., & McGrady, A., (eds.), (2001), *Media and the Marketplace – Ethical Perspectives,* (Dublin: Institute of Public Administration).
2. Cassidy, J. 2003 dot.con: the greatest story ever sold [sic] Harper Collins, New York, p. 21, reminds us that it was only 10 years ago [1993] that the University of Minnesota, in what was the first move to raise revenue from the internet, created a storm of protest when it imposed an annual license fee to non-academic users for its Gopher interface.
3. Feather, J., (2002), 'Theoretical Perspectives on the Information Society,' notes that, 'Perhaps we are too close to the phenomena that we are trying to understand. Even taking an historically informed approach, we are still looking at a society of which we are a part and in which we live and work.' in (Hornby & Clark: 2003, 15).
4. http://www.nua.ie/surveys/how_many_online/ NUA provides the following salutary caveat. 'The art of estimating how many are online throughout the world is an inexact one at best. Surveys abound, using all sorts of measurement parameters. However, from observing many of the published surveys over the last two years, here is an 'educated guess' as to how many are online worldwide as of September 2002. And the number is 605.60 million.'
5. Source: http://www.nua.ie/surveys/how_many_online/ (excepting Russia)
6. Telecomworldwire, 2/10/03 (LexisNexis) The Telecomworldwire gives Oftel, the British Internet regulatory body as the source for its information. The British Office for National Statistics, however, states that according to the February 2003 National Statistics Omnibus Survey, 50% of adults had accessed the internet in the month prior to the taking of the survey.
7. http://ccp.ucla.edu/pdf/UCLA-Internet-Report-Year-Three.pdf: p 18
8. cf. Review of Norris, Pippa, (Jan. 2003) 'Digital Divide: Civic

Engagement, Information Poverty, and the Internet Worldwide,' in, *Canadian Journal of Communication*, V. 28(1) (LexisNexis).
9. Source http://www.nua.com/ surveys/how_many_online/ africa.html. The date given for most of these figures is late 2001.
10. http://www.nua.com/surveys/how_many_online/s_america.html,
http://www.nua.com/surveys/how_many_online/africa.html and http://www.nua.com/surveys/how_many_online/asia.html. Data for Russia from Global News Wire, 18/03 (LexisNexis).
11. Nunberg, G, [undated] 'Will the Internet Always Speak English?' in, *The American Prospect,* V.11, #10. (http://www.prospect.org/print/V11/10/nunberg-g.html) Geert Lovink, among other commentators, has pointed to the development of different kinds of English on the Internet, including 'Euro English.' (Lovink: 2002, 123)
12. cf. UCLA Internet Project, Year Report; pgs 13, 54, 62, 68 & 79 (http://ccp.ucla.edu/pages/internet-report.asp)
13. Hotchkiss, G. (4/14/03) 'The Search Engine End Game.' Available from: http://www.promotiondata.com/article.php?sid=347
14. One estimate suggests that $6.2 billion dollars will be spent on Internet advertising in 2003:
(http://www.nua.ie/surveys/index.cgi?f=VS&art_id=905358749&rel=true)
15. Hotchkiss, G. (2/26/03) 'Adware & Spyware: Beware!' Available at http://www.searchengineposition.com/info/netprofit/spyware.asp
16. The National Archives (2002) Available at:
http://www4.nas.edu/onpi/webextra.nsf/44bf87db309563a0852566f2006d63bb/13a0fdabb8339dce85256bac005b4a2f?OpenDocument
17. cf.http://www.nua.com/surveys/index.cgi?f=VS&art_id= 905358207 &rel=true
18. http://www.nua.com/surveys/index.cgi?f=VS&art_id=905357766&rel=true
19. http://www.nua.com/surveys/index.cgi?f=VS&art_id=905357825&rel=true
20. http://www.nua.com/surveys/index.cgi?f=VS&art_id=905357136&rel=true
21. CNN: 'Live From…' 6/16/03; Transcript # 051613CN.V85 (LexisNexis).
22. Cited in (Cassidy: 2002, 21).

Bibliography

Burnett, Robert, & Marshall, P. David, (2003), *Web Theory: An Introduction,* (London: Routledge).

Cassidy, John, (2002), *dot.con: the greatest story ever sold,* (New York: HarperCollins).

Dhillon, Gurpreet, (2002), *Social Responsibility in the Information Age: Issues and Controversies*, (Hershey, PA: Idea Group Publishing).

Gerstner, Jr., Louis V., (2002), *Who Says Elephants Can't Dance?*, (New York: Harper Business).

Hornby, Susan, & Clark, Zoë, (eds.), (2003), *Challenge and Change in the Information Society*, (London: Facet Publishing).

Kenny, Charles, (Jan.-Feb. 2003), 'Development's False Divide: Giving Internet Access to the World's Poorest Will Cost a Lot and Accomplish Little,' in *Foreign Policy*, 76.

Lovink, Geert, (2002), *Dark Fiber: Tracking Critical Internet Culture*, (Cambridge, MA: The M.I.T. Press).

Mansell, Robert, (ed.), (2002), *Inside the Communication Revolution: Evolving Patterns of Social and Technical Interaction*, (Oxford: Oxford University Press).

Newman, Nathan, (2002), *net Loss: Internet Prophets, Private Profits and the Costs to Community*, (University Park, PA: The Pennsylvania State University Press).

Patvathamma, N., (March 2003), 'Digital Divide in India: Need for Correcting Urban Bias' in, *Information Technology and Libraries*, V.22, i1, p 35.

Preston, Paschal, (2001), *Reshaping Communications: Technology, Information and Social Change*, (London: Sage Publications).

Seebacher, Uwe. G., (2002), *Cyber Commerce Reframing: The End of Business Process Reengineering?*, (Berlin: Springer).

CHAPTER FIVE

Between Salvation and Destruction: On Heidegger's Thinking Concerning Technology

Jones Irwin

Introduction

By focusing on Heidegger's text *The Question Concerning Technology,*[1] this article will seek to address the implications of Heidegger's philosophy of technology. Heidegger's understanding of technology is based on a radical critique of the historical development of philosophy, in particular from the Roman period onwards. Heidegger offers a rich and persuasive re-reading of Greek philosophy, with especial reference to Aristotle's discussion of causality, but also invoking a more esoteric reading of Plato. Perhaps more importantly, Heidegger redirects these ancient philosophical resources to the crisis of twentieth-century humankind, pointing in the process towards the possibility of an exit from what he sees as our instrumentalist malaise.[2]

Heidegger's Thinking Concerning Technology

> The essence of technology is by no means anything technological (Heidegger: 1993, 311).

One of the major targets of *The Question Concerning Technology* is what David Farrell Krell has referred to as a 'technical technological' (Krell: 1993, 308) thinking, a system of thought which would seek to reduce all questions concerning technology to a merely technicist paradigm. Such a 'technical technological' philosophy views all questions concerning technology as *intra-technological,* that is as answerable only within a technical means-ends rationale. For Heidegger, this understanding of technology in means-ends instrumentalist terms (i.e. 'does it work?') only covers one aspect of the essence of technology:

> We ask the question concerning technology when we ask what it is. Everyone knows the two statements that answer our question. One says: Technology is a means to an end. The other says: Technology is a human activity. The two definitions of technology belong together' (Heidegger: 1993, 312).

From this point of view, while not completely inaccurate, the 'technical technological' understanding of technology is one-sided. Heidegger's meditation on the nature of technology thus seeks to broaden our understanding of what technology is, to extend the meaning of the 'essence of technology' (Heidegger: 1993, 311). In Heidegger's terms, the means-ends 'instrumental' (Heidegger: 1993, 313) definition of technology is 'correct' (Heidegger: 1993, 313) but not 'true' (Heidegger: 1993, 313):

> The correct always fixes upon something pertinent in whatever is under consideration. However, in order to be correct, this fixing by no means needs to uncover the thing in question in its essence. Only at the point where such an uncovering happens does the true propriate. For that reason the merely correct is not yet the true (Heidegger: 1993, 313).

Heidegger's analysis of technology in *The Question Concerning Technology* will thus seek to move beyond the merely 'correct' definition of technology to a 'true' understanding of the essence of technology. In order to direct this enquiry, one must ask questions which take us beyond the merely 'technical technological' (Krell: 1993, 308) level:

> The correct instrumental definition of technology still does not show us technology's essence. In order that we may arrive at this, or at least come close to it, we must seek the true by way of the correct. We must ask: What is the instrumental itself? Within what do such things as means and end belong? (Heidegger: 1993, 313).

Although technology is conventionally seen as a product of modernity, indeed an effect of the scientific revolution in the late eighteenth century (Krell, 309), Heidegger wishes us to consider an alternative definition. At the beginning of *The Question Concerning Technology* (Heidegger: 1993, 311), he warns the reader that a true questioning will lead to something 'extraordinary' (Heidegger: 1993, 311) in our understanding of technology. Krell succinctly describes this new consideration of technology:

> The advent of technology – and it is this historic, essential unfolding or provenance that Heidegger means by 'essence' – is something destined or sent our way long before the eighteenth century. One of Heidegger's most daring theses is that the essence of technology is prior to, and by no means a consequence of, the Scientific Revolution (Krell: 1993, 309).

This new description of technology is grounded in Heidegger's aforementioned claim that any definition of technology must operate on two levels: a) the sense that technology is a means-ends activity; b) the sense that technology is a human activity (Heidegger: 1993, 312). To the extent to which the latter is the case, one can also say that technology is a perennial activity i.e. it belongs to the essence of human culture itself from time immemorial. It is this conviction which leads Heidegger to consider the wider essence and nature of technology in the context of the important concept of 'causality'. Here Heidegger returns most especially to Aristotle's discussion of causality in the *Nicomachean Ethics* (and again in returning to an ancient Greek understanding, the implication is that technology is not merely a product of modernity) (Heidegger: 1993, 314ff).

But what is this link, for Heidegger, between the essence of technology and the ancient Aristotelian doctrine of causality? As traditionally defined (and inherited) by the history of philosophy, the Aristotelian doctrine of causality outlines four fundamental causes at the heart of human activity:

1) The material cause (or *causa materialis*), that is the matter out of which something is made, e.g. the material cause of a silver chalice is the silver matter.

2) The formal cause (or *causa formalis*), that is the form into which the material enters, e.g. the formal cause of a chalice is its shape.

3) The final cause (or *causa finalis*), that is the purpose for which something is made, e.g. the final cause of a chalice might be the rite in which the chalice is employed.

4) The efficient cause (or *causa efficiens*), that is that which generates or brings about the final product, e.g. the chalice is made by the silversmith (Heidegger: 1993, 313-314).

For Heidegger, the essence of technology can only be understood in the context of such causal processes. In other words, modern technology is nothing new but rather a contemporary expression of human causality at work in the world:

> What technology is, when represented as a means, discloses itself when we trace instrumentality back to fourfold causality. (Heidegger: 1993, 314)

In typical Heideggerian fashion, however, such 'disclosure' of technology, although moving beyond the 'obscure and groundless' (Heidegger: 1993, 314) contemporary definitions of technology, nonetheless only leads to a need for more questioning and thinking. If Aristotelian fourfold causality is the path to understanding technology, it is still a rather 'dark' path:

> But suppose that causality, for its part, is veiled in darkness with respect to what it is? ... From whence does it come that the causal character of the four causes is so unifiedly determined that they belong together (Heidegger: 1993, 314).

Having employed the traditional understanding of Aristotelian causality to expand the conception of technology, Heidegger now seeks to problematise this reading of Aristotle. Through a re-reading of Aristotelian causality (and reference to Plato's *Symposium*), Heidegger undertakes a more radical consideration of the meaning of both causality and, by implication, of technology.

Re-Reading Aristotle (and Plato) on Causality

Heidegger's fundamental disagreement with the traditional interpretation of Aristotelian causality is that it stresses one kind of causality as paradigmatic. According to Heidegger, philosophical thinking on causality has been led down a cul-de-sac by the almost exclusive emphasis on efficient causality, at the expense of the other three causes. This has also led to a fragmentation of the causes, when the original meaning of Aristotle was of a complete unity of causality:

> For a long time we have been accustomed to representing cause as that which brings something about. In this connection, to bring about means to obtain results, effects. The *causa efficiens*, but one among the four causes, sets the standard for all causality. This goes so far that we no longer even count the *causa finalis*, telic finality, as causality (Heidegger: 1993, 314).

Additionally, this interpretation of causality as primarily efficient causality has also had a rather severe effect on our understanding of technology. Although modern discourse has made the correct connection between technology and causality, it has read such a link as exclusively grounded in efficient causality, i.e. technology has been interpreted as a means-ends efficient causality. To the extent that Heidegger wishes to broaden our understanding of causality, he also thus wishes to broaden our understanding of technology.

This approach draws Heidegger back to a careful reading of both Aristotle and Plato and, in particular, an emphasis on the implications of their philosophies for a consideration of both causality and technology:

> But everything that later ages seek in Greek thought under the conception and rubric 'causality' in the realm of Greek thought and for Greek thought *per se* has simply nothing at all to do with bringing about and effecting. What we call cause (*Ursache*) and the Romans call *causa* is called *aition* by the Greeks, that to which something is indebted [*das, was ein anderes verschuldet*]. The causes are the ways, all belonging at once to each other, of being responsible for something else (Heidegger: 1993, 315).

Beginning with the Roman interpretation, therefore, and continuing up to the present day understanding, there has been, according to Heidegger, a gross misinterpretation of the complexity and unity of causality. In effect, this misinterpretation corresponds to what Heidegger has earlier described as a misinterpretation of the essence of technology (Heidegger: 1993, 312). In both cases, the emphasis on efficient causality has led to what Heidegger calls a 'correct' but not a 'true' understanding:

> The Greeks have the word *aletheia* for revealing. The Romans translate this with *veritas*. We say 'truth' and usually understand it as correctness of representation (Heidegger: 1993, 318).

Put simply, then, the problem for Heidegger is that since Roman times, the respective essences of causality and technology have been understood in terms of *veritas* or correctness, when they should have been understood in terms of *aletheia* or truth. Although sharing in this common misinterpretation, the cases of causality and technology are somewhat different, and Heidegger (through reference to both Aristotle and Plato) seeks to explicate their independence. Although the discussion of causality in this context is highly significant, I will only refer to it here insofar as it bears on our primary focus of the meaning of technology for Heidegger.

To clarify his meaning here, Heidegger makes a crucial distinction between a causality understood simply in terms of efficient causality and a more 'inclusive' (Heidegger: 1993, 316) understanding of causality. In the former case, causality is understood as simply 'that which brings something about' (Heidegger: 1993, 314) while in the latter case, causality is understood more broadly as 'that to which something is indebted'

(Heidegger: 1993, 314). Here, one is given the example of the silver chalice. Understood in terms of efficient causality, the cause of the chalice is defined as the silversmith whose role is interpreted as simply bringing the chalice into existence in the crudest sense (Heidegger: 1993, 315). Understood in contrast in terms of a more inclusive process, causality takes on a more complex meaning.

In the first case, although each of the traditional Aristotelian four causes play a part in the coming to be of the thing, nonetheless each of the causes is reinterpreted against the more instrumentalist understanding which Heidegger thinks has distorted their meaning. Thus, if we take for example the material cause, we cannot say that the 'silver' causes the chalice, but what we can say is that the chalice is 'indebted' to the silver without which it could not be. Similarly, in terms of the formal cause, the shape does not make the chalice but nonetheless without its shape the chalice would not subsist (again for Heidegger this is a relation of 'indebtedness': 1993, 315). This interpretation is repeated in terms of the respective final and efficient causes. Although the meaning of 'indebtedness' is enigmatic, nonetheless what is clear is that Heidegger wishes us to move away from an instrumentalist, means-ends understanding of causality. This is the main significance of the discussion for Heidegger's interpretation of technology. For Heidegger, causality is a process involving the four causes but additionally what amounts to a kind of fifth element, what he terms 'occasioning' (Heidegger: 1993, 316) or an 'inducing to go forward' [*Ver-an-lassen*] (Heidegger: 1993, 316). Another term Heidegger uses in this context is 'bringing forth' [*Her-vor-bringen*], and here he refers back to Plato's *Symposium*:

> Plato tells us what this bringing is in a sentence from the *Symposium* [205b]: 'Every occasion for whatever passes beyond the nonpresent and goes forward into presencing is *poiesis*, bringing forth [*Her-vor-bringen*] (Heidegger: 1993, 317).

Both Plato and Aristotle, then, understood causality as a fundamentally holistic process of a thing's complete coming-to-be or 'bringing forth', which included all four causes (understood non-instrumentally) in addition to a more fundamental condition of 'occasioning', which allowed the causes to be causes. That is, 'occasioning' appears to be Heidegger's term for the ontological ground of causality itself, the very 'becoming' of Being itself. This radical understanding of causality, which Heidegger understands as an authentic reading of Platonic-Aristotelian philosophy, has important implications when applied to the concept of technology.

Re-Reading Technology (Techne)

> Technology is therefore no mere means. Technology is a way of revealing (Heidegger: 1993, 318).

The above discussion of causality maps directly onto the consideration of the concept of technology in Heidegger, insofar as Heidegger is claiming that the misinterpretation of causality since the Romans is simultaneously a misinterpretation of 'the essence of technology'. Just as holistic causality has been reduced by modern discourse to the status of mere efficient causality, so too the modern understanding of technology as 'mere means' (Heidegger: 1993, 318) covers over a more holistic understanding of technology as 'a way of revealing' (Heidegger: 1993, 318). This misreading of technology is bound up for Heidegger with what Krell describes as the view of technology as a product of the eighteenth century scientific revolution (Krell: 1993, 309):

> 'It is said that modern technology is something incomparably different from all earlier technologies because it is based on modern physics as an exact science ... But it remains a mere historiological establishing of facts and says nothing about that in which this mutual relationship is grounded (Heidegger: 1993, 320).

Although it may be true that the modern examples of technology are somehow inextricable from modern developments in science, Heidegger appears to be claiming that this modernity of technology is nonetheless secondary to and dependent upon a more originary grounding of technology in Being, as a 'way of revealing'. Nonetheless, technology as a way of revealing is different in kind from what Heidegger as already referred to as a 'bringing forth'. Technology may be a way of revealing, but it is a specific kind of revealing, which carries its own dangers:

> And yet, the revealing that holds sway throughout modern technology does not unfold into a bringing-forth in the sense of *poiesis*. The revealing that rules in modern technology is a challenging [*Herausfordern*], which puts to nature the un-

> reasonable demand that it supply energy which can be extracted and stored as such (Heidegger: 1993, 320).

Perhaps the most evocative and powerful example which Heidegger supplies us here of the difference between 'bringing forth' and a more technological revealing relates to the development of the river Rhine as an industrial entity. On the side of a revealing which is an authentic 'bringing forth' of the Rhine, Heidegger cites two earlier examples. In the first case, the 'old wooden bridge' (Heidegger: 1993, 321) which joined one bank to another bank for hundreds of years. In another case, the poetic hymn 'The Rhine', as uttered by the German romantic writer Hölderlin. In both these cases, Heidegger's point is that an authentic causality is respected, where the river Rhine is allowed to 'bring forth' itself, in a process which allows it to authentically be.

However, modern technology 'reveals' the Rhine in an altogether more severe manner:

> In the context of the interlocking processes pertaining to the orderly disposition of electrical energy, even the Rhine itself appears to be something at our command. The hydroelectric plant is not built into the Rhine river as was the old wooden bridge that joined bank with bank for hundreds of years. Rather, the river is dammed up into the power plant (Heidegger: 1993, 321).

It is in this sense that Heidegger describes technology as a way of revealing which is a 'challenging' rather than merely a 'bringing forth'. At this point, Heidegger seeks to invent a new vocabulary to render the specificity of this process of technological revealing. This vocabulary is both intricate and difficult and a detailed examination of its nuances would take us beyond the limits of this particular essay. However, to conclude, I would like to focus on one particularly important term which Heidegger introduces here – *Ge-stell* [enframing] (Heidegger: 1993, 324). Characteristically, Heidegger offers the reader a rather enigmatic definition of this most crucial of concepts:

> [Ge-stell is] the challenging claim that gathers man with a view to ordering the self-revealing as standing-reserve (Heidegger: 1993, 324).

Unlike the more holistic revealing described by Heidegger as a 'bringing forth', *Ge-stell* rather describes the framework in which, to quote Bacon's famous phrase, man seeks to be a 'master and possessor of Nature'. Simply put, *Ge-stell* refers to the objectification and reification of nature, so evocatively exemplified by Heidegger in the case of the industrialised Rhine. To this extent, Heidegger would appear to be repeating the instrumentalist definition of technology which he has criticised as appearing with Roman philosophy. However, Heidegger's point here is that while technology in its applications may be objectifying, the process by which it objectifies cannot be understood simply in objective terms. That is, while we can explain *how* the Rhine is industrialised in technical terms, we cannot explain the *why* of its industrialisation in technical terms. This leads Heidegger to the ambiguous conclusion that the essence of technology is 'enframing' (Heidegger: 1993, 325) while being simultaneously enigmatic and 'mysterious' (Heidegger: 1993, 333). As Krell observes,

> As the essence of technology, enframing would be absolute. It would reduce man and beings to a sort of 'standing reserve' or stockpile in service to, and on call for, technological purposes. But enframing cannot overpower or even reveal its own historic, essential unfolding, nor indeed the advent, endurance, and departure of beings (Krell: 1993, 309).

For Heidegger, in conclusion, it is this mysterious 'destining' [*Geschick*] (Heidegger: 1993, 329) of technology which remains open-ended. It is clear that the cultural effects of 'enframing' have been devastating, not least upon humanity, 'In truth, however, precisely nowhere does man today any longer encounter himself, i.e. his essence' (Heidegger: 1993, 332).[3] Nonetheless, the future revelation of technology maintains a double possibility, of both destruction and healing. Invoking Hölderlin, Heidegger notes:

> But where danger is, grows the saving power also (Heidegger: 1993, 340).

To this extent, technology still holds out the possibility of the very 'bringing forth' and 'occasioning' which Heidegger so valued in Hellenic culture.[4]

Endnotes

1. 'The Question Concerning Technology' was originally delivered as a lecture by Heidegger as 'The Enframing' to the Bremen Club on December 1, 1949. It was completely revised and delivered as 'The Question Concerning Technology' to the Bremen Club on November 18, 1953.

2. It is noteworthy that 'The Question Concerning Technology' most definitely belongs to the work and thought of the later Heidegger, written as it is more than twenty years after Heidegger's magnum opus *Being and Time*. Many commentators point to a 'turn' (*kehre*) in Heidegger's work during the 1930s and to this extent, my analysis of this text is more applicable to the later than earlier Heidegger. For a lucid discussion of the 'kehre' cf. Richard Kearney, *Modern Movements in European Philosophy* (Manchester, Manchester University Press, 986).
3. For a significant critique of a residual anthropomorphism in Heidegger on this point cf. Jacques Derrida 'The Ends of Man' in *Margins of Philosophy* (Chicago, University of Chicago Press, 1982).
4. The question of the so-called 'destining' of technology and Being in Heidegger cannot, it seems to me, be addressed independently of Heidegger's interpretation (at least in the 1930s) of the 'destiny' of Nazism. For a balanced analysis of this issue cf. Hans Sluga, *Heidegger's Crisis: Philosophy and Politics in Nazi Germany* (Cambridge, Harvard University Press, 1993).

Bibliography

Aristotle (1976), *Ethics*, (London: Penguin).

Derrida, Jacques, 'The Ends of Man' in Bass, Alan (1982), translation with additional notes, *Margins of Philosophy*, (Chicago: University of Chicago Press).

Heidegger, Martin (1993), 'The Question Concerning Technology' in, Krell, David Farrell, *Basic Writings*, (London: Routledge).

Kearney, Richard (1986), *Modern Movements in European Philosophy*, (Manchester: Manchester University Press).

Krell, David Farrell (1993), *Basic Writings*, (London: Routledge).

Krell, David Farrell (1992), *Daimon Life: Heidegger and Life-Philosophy*, (Bloomington, IN: Indiana University Press).

Plato (1990), *The Symposium*, (London: Penguin).

Rosen, Stanley (1993), *The Question of Being: A Reversal of Heidegger*, (New Haven, CT: Yale University Press).

Sluga, Hans (1993), *Heidegger's Crisis: Philosophy and Politics in Nazi Germany*, (Cambridge, MA: Harvard University Press).

CHAPTER SIX

Tillich's Ontological Solution

Brian Donnelly

Paul Tillich is regarded as one of the outstanding and creative thinkers of the twentieth century. Referred to as a theologian of culture, he sought to establish the religious substance and relevancy in every circumstance no matter what the conflict or culture prevalence obtains, or the situation in which religion has seemingly been compromised or obviated. Tillich belongs to that small and elite group of Christian thinkers who have bravely engaged the modern world on its own terms and sought to restore the credibility of religion through the character and cultural forms that mediate our modern living and in which, and through which, our consciousness is bound.

But what does religion have to offer by way of explanation or instruction to our modern technological age? And is religion equipped and cognitively able to dialogue with scientific concerns that form the basis of our present-day technology? Is there room for, or even a necessity for, faith in circumstances that are technologically defined and pragmatically engineered? These and other such questions, arising from the nature of our cultural location, are questions to which Paul Tillich, throughout his life, endeavoured to offer a theological answer. Yet Tillich's solution was not an appeal to biblical revelation or a supernaturalistic interpretation of faith, but rather one of investigating the ontological basis of religion as being fundamental to all human concerns and achievements. For him the overcoming of every polarity and ambiguity in the human situation was a recognition of ontology as the essential structure of all facets of existence, including that of faith.

Religion is not a superimposition, a supernaturalism, but an occurrence of the human condition, inextricably bound and correlative to the phenomenology of existence. The explanation of religion is the explanation of existence. In this way, Tillich is an existentialist theologian. The existentialist questions are the perennial questions and must constitute the substance of any religious response. As Tillich reminds us, the human person is not merely an epistemological subject or an object of scientific inquiry, or technical management, but a being who must, behind the *sum* (I am) of Descartes' *Cogito ergo sum*, negotiate the human predicament of existence which is more than mere individual, subjective *cogitatio* (consciousness) (Tillich: 1952, 131). The starting point is the location of the human predicament and the manner by which the immediate, albeit ambiguous and finite situation, communicates the religious concern and points to that which is beyond and unconditional for the human person. This is the essence of religion – ultimate concern – and the endeavour by which this ultimate concern is brought to bear and finds expression in the structures of society, whether culturally, socially, politically or ecclesiastically, is the way by which religion is truly manifested. In other words, religion is implicated in all our situations no matter how we might describe or view that situation.

Ideologies and attitudes that deny ultimate concern and which by implication dismiss, denigrate or corrupt this religious intention can bring to bear a heteronomous culture. Tillich discovered this to his own personal cost. In 1933, because of his opposition to the dictatorship and heteronomy of Nazism and of Hitler's fascism, he was forced to flee his native Germany and to seek haven in the USA where he had to rebuild his life, work and reputation. What Nazism illustrated was the power and easy emergence of cultural situations where religion, rather than serving the ultimate concern, came to serve the ideological interest of race and nationhood. This exemplified for Tillich not only the vulnerability of religion, but also the courage and fight religion requires when facing possible corruption.

Religion is not in retreat nor should it acquiesce to what is simply popular or dominant in the cultural sphere. To that end there must exist a proper basis and use for the term 'religion', because, in the relationship between religion and culture, to facilitate its mediation, religion requires culture as culture itself requires an authenticity gained through an allegiance to religious substance and depth. As Tillich wrote:

> Religion as ultimate concern is the meaning-giving substance of culture, and culture is the totality of forms in which the basic concern of religion expresses itself. In abbreviation: re-

> ligion is the substance of culture, culture is the form of religion (Tillich: 1959, 42).

Similarly:

> The religious element in culture is the inexhaustible depth of a genuine creation. One may call it substance or the ground from which culture lives. It is the element of ultimacy which culture lacks in itself but to which it points (Tillich: 1968, 101).

The relationship between culture and religion is dialectic and interdependent. They must co-exist in the other to find their own relative expression. Accordingly, as Tillich sees it, the cultural forms by which religion actualises itself are those of language, art and other cognitive expositions that are heuristic and semiotic. These are the interpretative avenues by which there is an attempt to explicate the human predicament and to point to what lies beyond in terms of the ultimate. Technology pervades, then, these cultural forms and is constitutive of the very means by which ultimate concern is intimated.

Two aspects of Tillich's philosophy of religion seem pertinent and which bear responsibility to the consideration of technology and of the relationship between technology and of the religious element of transcendence. We speak here of the dynamics of faith and of the ambiguity of religion.

The Dynamics of Faith

'Faith is the state of being ultimately concerned, and the dynamics of faith are the dynamics of our ultimate concern', such is the opening remark of Tillich's classical work, *The Dynamics of Faith*. Ultimate concern seizes us in the manner of Rudolf Otto's *mysterium tremendum et fascinosum*. It is a presence or stirring which affects us. It is not something which we conjecture or fantastically conjure with our mind, but rather is a response to an awareness of a total Otherness which is, by definition, beyond objectification. It is not an object of apprehension but that which apprehends and grips us. As Tillich observes, for human beings this religious in-breaking lies at the heart of the meaning of our existence and correlates with our fundamental and persistent ontological quest for the ultimate and unconditioned reality. So, the manner by which we respond to the ultimate concern is the manner that determines how religion is borne in our lives. Ontology is inescapably religious and religion must concede to ontological. Tillich stated this principle in the following way: 'we are immediately aware of something unconditional which is the prius of the separation and interaction of subject and object, theoretically as well as practically' (Tillich: 1959, 22). Religion reflects the inner ontological propensity, Tillich insists, of the finite realm drawn to the infinite and where the human being is the locus of consciousness in this relation of the finite to the infinite.

Thus when the external expression of this interior ontological truth does not reflect the movement from the finite to the infinite, or of the conditional to the unconditional, true religion and the dynamic of faith fails to be exhibited. Similarly, an aberration of religion occurs when a finite object or a situation is elevated to that of the ultimate or the infinite, or faith allegiance claimed to that which is really a preliminary concern. This distortion occurs when denigrating or suppressing the awareness of the ontological truth that lies at the heart of human existence. The supplanting of the infinite and ultimate with a finite or preliminary concern has affected organised and institutionalised religions. It is an aberration when the religious character becomes a perversion of self-interest and self-regard rather than a conduit or instrument by which ultimate concern is mediated. Whether done wittingly or unwittingly, religion is diminished in its formal setting, becoming a travesty of what it ought to project and reflect.[1]

In addition, as Tillich observes, because the dynamic of faith is not the preserve of institutionalised religion solely, faith can be fostered, facilitated and exemplified by non-religious structures and secular forms and groupings. Ultimate concern is not confined to the mediation within what is traditionally viewed as institutionalised, formal 'religion'. In fact, as Tillich makes clear, ultimate concern, by its very definition of being unconditional and absolute, transcends and scrutinises every formal and so-called religious sphere, for the ultimate is unconditional and unrestricted in the possibility of being demonstrated in any or every circumstance of the human situation or spectrum of experience. Any so-called religious claim to absoluteness by an institution, church or congregation, can only be a claim at best to witness to the absolute and ultimate, in a relative and provisional manner, and not to be equated with absolutism itself.

This is an instructive insight for Tillich and one which places the relevancy of faith as being central to the interaction of religion and secular concerns and of how religion points, dialectically, to the ontological criterion by which cultural forms are measured and adjudged. But paradoxically, religion in itself is not without ambiguity and here lies the second important conceptual tool for Tillich and one which helps elucidate further the analysis we want to present concerning religion and technology.

The Ambiguity of Religion

A second tenet for Tillich is that all religious statements are symbolic. Religion can only be explained symbolically since the very norms of religion are conditioned. Yet, as we have mentioned, in its conditionality it can point towards the unconditioned. This is the essence of religion and it can reside both in formal and organised religious societies and churches, and in secular groupings and cultural forms. In its representation of the ultimate, religion seeks symbols so that even language and the very term 'God' is symbolic of Ultimate Being. Critically and starkly for Tillich the only non-symbolic statement is the religious assertion that the 'Ultimate being is Being-itself'. This exercises an important pre-eminence for Tillich, because outside of this lies the quagmire of religious symbolism: of their human invention, distortion, ambiguity and limitation.

Divine ontology means the non-objectification of God. Ontology is fundamental and prior to the subject-object distinction that makes thinking and accessibility to knowledge possible. Being and Being-itself transcends both objectivity and subjectivity. Yet, linguistically the term 'God' implies objectification and as Tillich argues, the seriousness of the Ultimate question is not to regard God as an object among objects. God transcends all objects and subjects and is above and prior to the subject-object cognitive division conveyed in all terminology and in the term 'God'. By being the Ultimate, and Being-itself, God is above the name God. This has resonance with Meister Eckhart's belief: 'I pray to God to rid me of God, because the word 'God' is an ambiguous, human invention.' So, the recognition of the God above God, of Being-itself, illustrates the ontological basis on which Tillich grounds his explanation of religion. But it indicates how the business of religion remains trammelled in spite of its aspirations – forced to concede to cognitive finitude. Tillich speaks of the self-transcendence of religion or of self-transcendence as the ontological character of religion. Any entity or manner of being that seeks self-transcendence engages in the religious task of ultimate concern-ing. This self-transcendence is a function of life itself. Yet there exists the counter force to life and to religion which resists self-transcendence. This Tillich named as a profanisation or a demonisation[2] which, while evident in the dynamic of religion, denotes the very ambiguity that affects existence itself in the way that life, transcending itself, remains at the same time within itself.

Importantly, Tillich speaks of the *self*-transcendence rather than simply the transcendence of being, meaning that within itself, the finite entity points beyond itself, and that underpinning this important dynamic, religion is precisely this paradox of pointing beyond what lies within; of the correlation between the infinite and the finite. Transcendence can denote a fleeing from the sphere of the concrete and historical, and of decision-making, to a world of ideal forms or that of 'super-nature'. But transcendence cannot be an escape from present reality or concerns. For a being to be self-transcending or to be given a 'sense' of ultimacy it must be essentially finite. As Tillich posits, the task of religion is to be engaged with the existentiality of human finitude and not to prescribe or foist a supernaturalism that obviates or dissociates religion from the level of the finite (Tillich: 1968, 69). This is how he clarified his position:

> The transcendentalising act does not signify that we possess the transcendental. The point is that we are in quest of it. But on the other hand this quest is possible only because the transcendental has already dragged us out beyond ourselves as we have received answers which drive us to the quest. The development of this dialectic is the proper aim of philosophy of religion and of the improperly so-called 'natural theology'. Only it should not be supposed that this can be a substitute for the theology of revelation (Tillich: 1935, 140).

By indicating something of how Tillich viewed the dynamic of faith and of what he argued was the ambiguity of religion, this existential-ontological analysis forms the elementary basis of how Tillich approached the religious concerns surrounding modern technology.

Technology

All of nature acts technically, the term *techne* coming from the Greek meaning art, skill, or cunning of hand. Nature has an innate cunning to achieve, to survive and to be. The *logos* that is brought to *techne*, which we describe as technology, bespeaks the reasoning and human rationale brought to bear upon nature's cunning; the designing, manipulative and calculating capacity of the human mind. But in co-operation with human reasoning, there is present the values and interests which motivate and determine the form and shape of human involvement. This influences the aim, achievement and purpose of technology. In the modern context, since the nineteenth century, this has been greatly determined by the advances in the physical and natural sciences.

Already we can see the decisive role ultimate concern can play. Unconditional purposefulness results in the unconditional rationality of the technological structure. This has significant implications upon cultural forces such as the market economics of the consumerist society. Here the rationality that is inclined to condition technology is one of exploitation and greed. Technology's corruption to capitalist economics is an area of huge interest and one which Tillich explored through his Marxist interpretation of religion.

Tillich named what he termed four cultural symptoms that arise from modern technological thinking and which in turn pose religious difficulties. The symptoms he identified were:

1. The inclination of technological thinking to remain at a superficial, 'horizontal' level of living.
2. The intention to control nature.
3. The overriding perception of seeing everything as calculable objects.
4. The reduction of reason to what one could call 'calculating' or 'technical reasoning'.

Tillich regarded the ascendancy of technical reasoning as a devastating assault on reasoning itself. The restorative solution is one of reconditioning reason by a redirection through an ontological reference. This reveals the primacy of ontology and the survival of the transcendent element of religion against the reductive tendencies of technical reasoning. He wrote:

> Reason was not always the tool simply of the businessman, the technician, or of scientific analysis. It was formerly the power of knowing the ultimate principles of the good, the true and the beautiful ... It has been replaced by something which in former centuries was only secondary: humanity's rational power for control (Tillich: 1996, 26).

Technological reasoning then needs to be ontologically informed to recognise the dimension of being. Technology is a means to an end, not an end in itself. And while the incentive of technology is production, inherent to technology is nature's creative will, its purposefulness of both the whole and its parts, and its inner, necessary drive towards the ultimate. The vital and accomplishing role of the *logos* in technology requires it to be open to the in-breaking of the ultimate concern. This is the only reliable and objective criterion by which to gauge and measure the purpose and value of the rational conditioning that surrounds technology and of its consequent culture. Otherwise humanity is at the mercy of unchecked influences increasingly liable to demoralise and dehumanise.

Technology provides newer patterns or expressions of living and of cultural contexts through which ultimate concern is to be uncovered and interpreted. As Tillich notes, technology can free humans from the unrelenting stress of bodily pain, and the stifling oppression of daily evils, but it also projects the power of rationality that can exploit, dehumanise and objectify the human person and cause the loss of ultimate concern. The task for humanity is to appropriate technology according to the primary and fundamental concern of that which is ultimate, unconditioned and absolute.

> We must incorporate technology into the ultimate meaning of life, knowing well that if technology is godlike, if it is creative, if it is liberating, it is still also demonic, enslaving, and destructive. It is ambiguous as is everything that is: not more ambiguous than pure spirit, not more ambiguous than nature, but as ambiguous as we are (Tillich: 1975, 302).

As mentioned earlier, the closure of reason to the ultimate results in the loss of religion. This has been further described by Tillich, in typical existentialist philosophy, as the domination of the horizontal dimension over the vertical dimension of 'depth'. The loss of the depth mirrors the loss of the self, where the human situation becomes ever more precarious and vulnerable to estrangement and emptiness. For Tillich the 'fight' against technology is not against technology *per se* but what arises from technology in terms of the dehumanising effects, where technology can become a structure of objectivisation, transforming life and person into a thing, an object of calculation and exploitation. This technicalism, as Tillich names it, like scientism of science, is a violation and loss of the depth dimension. It becomes not just a violation of human concerns and of ultimate concern, but it is also a violation of technology itself. The purpose and meaning of technology must reflect its inner aim and claim of being which is an extension of Being-itself. Technology has its own latent religious element that needs to surface and be realised. No area of human creativity or cultural form is devoid of religious possibility, as Tillich claims. So there exists also within technology that power to counter the de-humanisation and profaning influence of technicalism. As Tillich argues consistently, the religious and the secular are not separate realms but are within each other. What secures and rescues technology from technicalism is the ontological reference to the ultimate or what Tillich has also described as self-transcendence; that technology, like religion, is potentially a medium which can point to what is ultimate and beyond itself:

> Never consider the secular realm Godless just because it does not speak of God. To speak of a realm of divine creation and providence as Godless is Godless. It denies God's power over the world. It would force God to confine Godself to religion and church (Tillich: 1996, 62).

How Persuasive is Tillich's Argument?

Tillich explains the phenomenon of religion from the basis of ontology and thereby infers that there is a point of identity or possible synthesis between the a-personal, ontological description, Being-itself, and the personalism of God in biblical faith. But is this coherently and persuasively argued by Tillich, or is his ontological abstraction, while a rehabilitation of religion is a distortion and a detraction of the meaning of faith? Faith requires a more personable reference to God than the concept Being-itself. Moreover, there is no such thing as a saving ontology in the way one might think of a saving faith. Ontology cannot be a substitute for faith, even though it can be a means by which the cultural is prised open to regard the existential concerns that access faith. Unlike Heidegger, Tillich sees ontology as leading towards a theistic Being-itself. This is not simply a philosophical abstraction for Tillich but an important explanation of faith, and significant also when anchoring religion within the faith community or church. The personalism of God and the ontological description of ultimate concern must coalesce. Tillich wrote:

> Being and person are not contradictory concepts. Being includes personal being; it does not deny it. The ground of being is the ground of personal being, not its negation (Tillich: 1955, 83).

To those who hold that theology is *theo*-logy and not *onto*-theology, Tillich would argue that all systems of thought are sourced in ontology. It is prior to epistemology. Any theology or philosophy of religion gains credibility by speaking about and speaking through the burning issues of the day, which ultimately bespeaks of the existentialist-ontological reference. Unlike Karl Barth, who held all theology accountable solely to the biblical word, Tillich saw it a matter of a two-way accountability; the human world and the biblical word are in dialogue and consequently culture is a necessary and unavoidable condition through which faith is reasoned and proclaimed. He declared that 'if one starts to think about the meaning of biblical symbols, and the incorporation of language and literary forms, one is already in the midst of ontological problems (Tillich: 1955, 83).

Of course there are many thinkers like Barth who are revelation-centred, non-correlational and anti-ontological, who argue that human beings are incapable of disclosing God independent of the deliberate condescension of God's self-revelation. For them God stands beyond human possibilities and there is an absolute distance between the infinite and the finite which humankind is unable to cross without divine intervention. Humankind has no inherent ability to perceive God and so we obtain an understanding of God not through any efforts of our own, such as that of an ontological inference, but by the means only of divine revelation.

Moreover, Tillich's faces criticism from others who claim that by adhering to an assumption that God can be understood in terms of Being or Being-Itself, implies that Being has a 'priority' over God. Being, as a metaphysical category, is a confinement of God, as is argued by those who regard that 'God' is above Being-itself or without Being. Perhaps it is better to say that Being-itself is an icon of God, not an equivalence of God (Marion). This has implications for faith and religion, as Heidegger conjectured, 'faith does not need the thought of Being. When faith has recourse to this thought, it is no longer faith' (Heidegger: 1980, 60).

Again, it could be argued that the term 'ultimate concern' cannot escape the subject-object relation. 'Ultimate concern' pushes out the linguistic boundaries but it still must be objectified even though it wishes to be non-objective. Language is beheld by this ontic structure of the division between subject and object. No matter how much Tillich points beyond this structure he must return to it. His language lapses back to what it longs to escape from or argue against. 'Ultimate concern' remains an objectificaton and in that sense, God remains objectified.

Tillich straddles both theology and philosophy and fails to gain full membership of either camp. Theologians criticise him for having built his theology in an ontological manner, while the philosophers reproach him for having solved philosophical problems with the help of theological ideas. But in spite of the limits of his thinking, and where he might have fallen short, the remarkable contribution Tillich makes is the great scrutiny and judgement he applies to both religion and culture. The method he adopted, by way of an ontological-existentialist analysis, may not be to everyone's liking and particularly for traditional theologians who are for the most part kerygmatic in their approach, but Tillich presents a body of thought worthy of attention and interest for both his supporters and critics alike (Bulman & Parrella: 2001). The issues that afflict religion are those that also afflict culture: ambiguity, vulnerability to profanisation and self-enclosed reasoning. Without the formidable intellectual legacy and path-finding achievements of thinkers, such as Tillich, religion may find few convinced and convincing friends left on its side in the twenty-first century as it does battle to gain credibility against the growing ascendancy and approbation of various cultural forms and attitudes, insidiously profaning civilisation by the dehumanising functionalism of technical reasoning.

Endnotes

1. Sadly, this tendency is evident in the churches at certain points of history, when the religious criterion of ultimate concern was usurped by more immediate and finite concerns of power, institutional wealth and privilege. So the self-relinquishing and self-negating character of religion is critical to its integrity and credibility.
2. Demon or demonisation refers to the Greek mythological figure *daimon* – a supernatural being of an intermediate nature between that of the gods and humankind. This is the sense and meaning used by Tillich.

Bibliography

Heidegger, Martin(1980), 'Séminaire de Zurich', trans., Staatdjian, D. & Féder in, *Po&Sie,* No.13, pp. 52-62.
Marion, Jean-Luc (1991), *God without Being* , trans., Carlson, Thomas A., (Chicago: University of Chicago Press).
Tillich, Paul (1952), *The Courage to Be,* (London: Fontana Press).
Tillich, Paul (1955), *Biblical Religion and the Search for Ultimate Reality,* (Chicago: University of Chicago Press).
Tillich, Paul (1959), *The Theology of Culture,* (Oxford: Oxford University Press).
Tillich, Paul (1968), *Systematic Theology* Vol. 3, (London: Nisbet Press).
Tillich, Paul (1996), *The Irrelevance and Relevance of the Christian Message,* (Cleveland, OH: Pilgrim Press).
Tillich, Paul (1968), *Gesammelte Werke* IX [Die Religiöse Substanz der Kultur], Albrecht R., *et al* (eds.), (Stuttgart: Evangelisches Verlagswerk).

CHAPTER SEVEN

Technology as Monumental History

Fiachra Long

I want to suggest an argument that proposes a serious link between technology and history. This is largely based on a hunch that as a society we need to know the historical importance of technology in order to assess its human value. But the key ideas I will use to develop this argument are the classical Aristotelian understanding of *techne* and the more modern understanding of monumental history adumbrated by Walter Benjamin but elaborated in a different way in Heidegger's *Question Concerning Technology*. These latter ideas have made their appearance in skirmish-like fashion during the twentieth century often either as a gloss on or as a criticism of Nietzsche's superman – the *Übermensch*. It still remains to be established what role Nietzsche's thought plays in understanding the function of technology in the world today where *techne* has come to represent human power and ingenuity but where this power is frittering away incrementally due to the impact of obsolescence. Indeed it is to be hoped that the description of a phenomenon of technology as monumental history might enable us to make some preliminary reflections on the narcotising influence of obsolescence in western culture today.

We begin in the middle where we usually begin. Signs of trouble have been emerging for a long time now and in various ways – the inability of the computer industry to hold on to what is the state of the art technology, the displacement in popular music of those who create music by those who manipulate sounds, sample them and republish these arrangements against the background of pre-recorded metronomic beats, the postmodern engendered disbelief in contemporary architecture where older styles conceal modern interiors or where public monuments are designed as self-effacing mirrors; the taste for retro in fashion and music which generates an appetite for representation over presentation; a general cynicism about appearance as if this age has had to settle for self-eclipse as a way of appearance. Speed has obviously contributed to this state of affairs because a culture travelling forward at speed is often not aware of its own imbalance.

An old science fiction series from the 1950s called *The Twilight Zone* provides an interesting metaphor for this situation. According to this scenario, a warp has occurred in space through which the unfortunate victim has unknowingly slipped only to find himself in another dimension or reality – the twilight zone. The high drama of this scenario is how the victim can find his way back to the real world. There are difficulties because victims seem to have lost all sense of where they are or how they are to manoeuvre. They do not know their left hand side from their right hand side and are unable to recognise anything from which to take their bearings. Happenstance might bring bodies together and they can cling to one another, each one as lost as the next until a voice from the real world provides them with possible bearings. The question uppermost in everyone's mind watching the show is the dramatic search to find the warp and discover thereby a way back through it to safety. The problem is that this hole is fluctuating and may close at any moment. To add to the tension, it keeps changing position even from the perspective of those living in the real world, and it often gets smaller, making every passing moment critical to the survival of those trapped. Will the victim get back or be lost forever? This was the problem for the television show. Our problem may well be similar, namely, that without monuments we are entirely lost, without the ability of *techne* to provide us with fixed co-ordinates marking out our time and space of life together we will not be able to find our way back from virtual worlds to reality itself.

If, however, one were to link technology not with the future but with the past, then the problem of technology would arise less from technology itself and more from the way technology fits into culture, or rather the way culture wraps itself around it. Invent a weapon of mass destruction nowadays and a culture soon develops to justify it and envelop it, hiding it from view or positing it as indispensable. The problems flow from the way technology is casually understood in our culture and by the way intellectuals align themselves on the issue. What guidance then is offered Wisdom's declaration that everything has been created by number, measure and weight? (*sed omnia in mensura, et numero, et pondere disposuisti* Wis XI, 21) Without such a balance, and given its tendency to serve as a base for a cultural wrap-around, technology may now be entering a self-destructive phase, that the pace of change may be making it impossible to recognise the vast cathedral in which its artefacts appear. My concern is basically to reinforce technology's links with the past; to situate it in its context as part of the created world and as an important part of the human cultural world.

Indeed it should be possible to think of technology in terms of artefacts that can be left behind in time. The skilled cabinet-maker leaves behind products which endure beyond the moment of production and last longer sometimes even than the life of the cabinet-maker. The skilled theatre actor creates an effect in the audience which will not be forgotten long after she has left the stage. Modern gadgets too, those reproduced digitally as well as those reproduced mechanically, retain something of this quality: they leave something behind that has a lasting effect on people. It seems essential in modern economies to create a product which will last long enough to count as a marketable good in the normal context of buying and selling and to resist obsolescence for as long as possible. But even in pre-capitalist cultures people have used artefacts to resist the forgetfulness prevalent in human experience concerning friends, heroes and actions and the general dissipation of one's life energy into disconnected and unrepeatable moments. From the cave drawings in Lascaux to the sketches of Blake, from the crafting of a Stradivarius to the design and construction of the Twin Towers, these artefacts have served as stability-markers against which individual lives as well as human culture have taken their bearings. The reverse also must be true for with the destruction of the 1500 year old Bamiyan statues by the Taliban, coupled with the destruction of the Twin Towers, any stable context for the appearance of technology has been compromised in the awareness of people and the populations of the world have been subjected to the fear that has for a long time been a feature of the western world, namely, that all human culture, as represented in its artefacts, is already obsolete.

As a general rule this suspicion reflects the fact that the products of technology are losing their historical significance in the secular sphere in much the same way that history itself has lost its significance for religious fundamentalists. This state of affairs casts into sharper relief our dependence on technical products to develop human culture and questions our need to respect the relationship between old products and young products, old cultures and new cultures, old values and new values, indeed the old and the young. If obsolescence or the denial of history ever intervened as a general rule of culture to reverse this relation then not only would human culture struggle to make an impact on its time and place, but technical reason also would lose its rationale. Technology would then tend to be associated with currency and scientism and indeed this may be what is happening today. Indeed today we see all the signs that technology itself is disappearing into a culture of obsolescence. The rapidity of industrial development in a digital age signals a much shorter life-span for products and because fashion is notoriously fickle, because artefacts almost immediately become dated and irrelevant, this creates a wobble in culture which affects the age old relationship between the young and the old and where for the first time in history the young, basically through technology, become the teachers of the old (Tapscott: 1998).

On a political level, if we look at ways in which post-WWII philosophers have been at pains to call for new beginnings, quite reminiscent of the reaction following the Thirty Year's War (1618-1648) and the rise of Cartesianism, it suggests that the Jewish holocaust has finally brought rational culture to its critical crossroads. The holocaust has undermined the entire Enlightenment belief in the dominant role of rationality in the behaviour of humans. For some this shock has taken the form of a note of warning to any political optimism, while Hannah Arendt pointed to a possible crisis in education (itself a political science) if members of the new generation were not able to express their own novelty, their own natality in a culture that must perforce be conservative (Arendt: 1977, 193). Arendt noticed the naïve turn towards novelty in Nietzsche's appeal for the re-evaluation of all values and she tried to argue for a mitigated kind of novelty, remaining critical of the naïve support for the progressivism pervasive in the US education system. She knew that most of the dangers implied in the concept novelty could be summed up in the desire to be entirely contemporary and that to hold on to the present moment as one's divinely gifted place would be to claim divinity for oneself after a fashion. And yet this trend is powerful and indefatigable. The desire to be eternally present-tense drives the consumerist appetite to grasp at state of the art technology before it disappears. Technology then becomes an unknowing accomplice in this quest for immortality. For reasons such as these I will argue here that technology needs to be conservative; it needs to be a science of the past, a marker for historical time. Here we come close to the paradox of the problem well-recognised by Arendt – that in order to be present one has to be in some sense in touch with the past. Technology can only bridge the past and the future by retaining a monumental quality because this is the quality it has always used to etch its mark on history. In the absence of such monuments, we no longer know who we are because we cannot see ourselves mirrored in the products we create.

To complicate matters, the phenomenon I describe here as the historical monumentality of technology is now disintegrating. So while I argue that traditional technology should therefore be less dedicated to the future than it is to the past, I recognise that this particular dedication comes under serious challenge in a consumer environment which makes its sales on the basis of products that present themselves as up to date and contemporary. This tension needs to be clarified and indeed this argument would become clearer if we looked more closely at two elements of this phenomenon – *techne* as history and *techne* as monument to discover how to read our present situation as it is already beginning to disintegrate. For the first concept we turn to Aristotle and for the second mostly to Heidegger.

1. Aristotle: Techne as History

In Aristotle's theory of ethics where an attempt is made to describe the legislative codes implied in moral action, truth is a motion of the mind. The idea of motion is important in Aristotle's account. Thought of itself, he says, moves nothing (NE 1139a) and yet, wedded to the desire for a *telos* or end, a motion is caused in the mind and this motion is what determines the type of truth apperceived by the mind. Aristotle identified five kinds of motion and hence five kinds of truth, each reflecting a different kind of motion in the mind. In each case the mind identifies a goal to be reached and ponders on the means to be used in achieving it. Each motion is teletic in structure and in each case the motion is measured with reference to a stable feature. Motion requires this kind of stability for it is against this that the motion can be evaluated. The five motions which Aristotle calls intellectual virtues or *aretai* are expressed best by the Greek terms *episteme* (scientific knowledge), *techne* (technical knowledge) *phronesis* (practical wisdom), *nous* (intuitive understanding) and *sophia* (intellectual wisdom). For the purpose of this argument, let us focus essentially on the first three dispositions in so far as each of them signals a simple motion in the mind and thus a temporal operation of thought while the actions of *nous* and *sophia* are too layered and complex to be evaluated here.

Episteme or scientific knowledge conceives of things that cannot vary, things like stars or planets. Because these things endure and outlast the life of the knower, they stand clear of the investigator's puzzlement and lack of knowledge. Knowledge is therefore a kind of motion in the mind terminating in a 'conviction arrived at in a certain way' (NE vi,iii,1139b30). Stars and planets stand there majestically inviting inquiry. Two motions are involved to further the inquiry. The first may be the way of induction (*epagoge*); the second by deduction (*sullogisme*). Thus in a deductive sequence, the mind is aided to assent to the truth of a motion linking premise to conclusion and if this motion is consistent with the laws of inference, the mind must assent to the conclusion in order to operate rationally. All S is P, but X is S, therefore X is P – All blue stars burn helium, but x is a blue star, therefore x burns helium or All French men are passionate, Pierre is a French man, therefore Pierre is passionate. If we have problems with this it is because we do not accept the premise – we may hold that all blue stars do not burn helium or that all French men are not passionate. It is not because the movement of the mind to assent to the conclusion is irrational. Indeed to prevent this movement from happening as long as one accepts the premise would be rationally impossible. Similarly if an inductive sequence respects logical rules, the rational mind moves automatically to assent to the premise on the basis of the evidence presented. This inductive movement is generally much less certain than logically valid deductive inferences and has been the subject of much debate over the centuries. Whether premises can be verified by such a process, or only falsified as Karl Popper maintained, is a matter for another debate. In either case motion is involved which is then to be tested by reference to a stability in the mind. This is the key point.

One should not be too surprised to learn that an epistemic movement of the mind in deduction relies on premises left behind, as it were, for the duration of the deductive process whereas in induction it is the experiment that is left behind in the search for suitable premises. These premises may only be challenged after a process of inquiry leading back from observed facts (which for a moment at least must remain uncontested and stable). Accordingly, if principles are not held with too much conviction, then an inductive process is needed to restore them. Overall, however, it is the premises or principles which serve as lynchpins for this mental movement, and since most Greeks believed that the gods knew things thanks to the stability of knowing principles, knowledge that emulated this movement could be considered quasi divine. Furthermore all scientific knowledge could be portrayed as concerning necessary objects whose behaviours are predictable in advance. If I know about trees in an epistemic sense, I know something general about them and this knowledge if correct has an apodictic or necessary quality and will not fluctuate as I observe one tree after another. Aristotle's *episteme* enjoys this apodictic quality which determines the type of stability the knowledge has and effects in turn the quality underlying any teaching of *episteme*. Whatever knowledge there is about objects can be communicated through teaching (NE vi, iii, 3 1139b 25) which involves setting up a motion in the mind of the learner that replicates the form of motion in the mind of the teacher. In this case the motion is circular from premise to conclusion and back again.

The important point to remember is that this mental process plays on a certain type of historicity. The historicity of *episteme* mimics 'eternal' knowledge after a fashion in so far as it claims to be valid for all time. So while the knowledge has been derived from an exercise of knowing that is temporal and discursive, the result is a kind of knowledge that claims to be able to withstand the test of time and to be in a sense eternal. There is a clear paradox here, for the way epistemic knowledge has been generated in the human mind is not the way it will be exercised and this is why for Aristotle such knowledge is close to the divine. Indeed ahistoricity is deemed to be an important quality in epistemic knowledge because experimental results need to be repeatable and tested by others.

The association with the divine order is not appropriate for the second kind of motion evident in the mind. In the case of *techne,* the mind's anchor is not only a start off point for knowledge but is continually renewing itself on the basis of the practical exercise of that knowledge. *Techne* is both practical and theoretical; it is something one has and it is something one does; it is an enabling kind of knowledge very much linked to events that might well be otherwise. Just to say that one knows how to ride a bike because one could do it as a child is no guarantee of being able to do so as an adult, and the accounts of people who have forgotten how to speak a language they once knew are legionary. So it is clear that the knowledge of *techne* because of its regulatory role in governing action (Aristotle insists that the word should be production not action) is vulnerable to time in a way that is quite different from the knowledge of *episteme* (which may of course be theoretically forgotten) and this is the point we need to explore further with Aristotle's help. Technical knowledge provides an insertion into history which epistemic knowledge does not provide. It leaves traces which have a public meaning that is both difficult to agree on and difficult also to forget.

For Aristotle the artist or craftsman produces something which does not exist of necessity and which might have been otherwise. The very contingency of a product separates it from an object of knowledge such as a star or a logical premise. Furthermore, the product might be very visible such as in the case of a wood carving or engraving and it might also be invisible in the case of habitual skill residing in the wood-turner's soul. The historical and technical aspect of the technical product is revealed as the wood-turner leaves a wooden bowl behind, the painter a painting, the actor a performance in the memory of those who were its witnesses. In all these cases something has happened to alter the course of history. This results in a type of historical learning which can have many consequences both for individuals and for human culture. For instance, the harpist produces musical sounds as she learns how to play the instrument and she can use this residue, the tune itself and the memory of the tune, to measure her skill as a player or to mark a cultural event and make an occasion memorable. The product which now may be the invisible performance of the harpist leaves an audible trace which now can be preserved on tape and by other means although for Benjamin this process, the mechanical reproduction of art, marks the beginnings of a decline (Benjamin: 1970). This is not to say that the historical representation of the work of art is also a sign of decline. Quite the contrary, for as Benjamin states:

> The authenticity of a thing is the essence of all that is transmissible from its beginning, ranging from its substantive duration to its testimony to the history which it has experienced (Benjamin: 1970, 223)

The presence of a work of art for Benjamin is not its epiphany at a fleeting moment but rather a collective meeting point where the many deposits of historical time come together to offer themselves to public gaze. The authority of technical work derives from its duration which in Benjamin's view can only be celebrated by the actualisation of history in ritual. Benjamin's position is a further elaboration of Aristotle's view that the performer's own development can be chronicled by means of an historical product which identifies not only the production of artefacts but also the spiritual changes which occur in the soul of an audience and a performer. Without a succession of such residues and without an ever more complex use of residual skills, the development of technical knowledge would not be identifiable. We are getting here the first clues as to the link between technical products and ordinary historical life. This is because the expression of technical mastery is manifested in two historical ways, both in the artefact that stands outside the production process as a historical momento and as a historical state of soul which remains capable of development or disintegration. In keeping with this second sense, Aristotle insists that *techne* is a quality or disposition (*hexis*) of mind or knowledge which operates according to the truthful art of technical thinking (*meta logou alethous poietike* (NE vi, iv, 4, 1140a 11) Technical thinking is true as a motion of the mind to the extent not only that it brings into existence an artefact which might not otherwise exist but also as a state of soul which is perfectible through the exercise of the art. If time were suspended it would seem that there was a circular relation between the soul of the harpist and the production of musical notes, but artefacts basically remain behind while the soul in a sense enlarges in experience, to use a spatial metaphor, although it does not necessarily improve in performance as a result of the exercise of the art. Good singers in middle life may be poor singers when they become aged. So this process is not simply circular. It is not like the theory of science, which remains perfect no matter how often it is applied. The truthful movement of *techne* takes its starting point from the residue left behind to reflect it in historical time and not only produces an artefact but generates something new and unexpected in the historical being of the technician. Indeed this learning process is based squarely on a production of some kind – the production indeed through the technical elaboration of human culture itself.

Given the fact that an artefact has been brought into being using an element of luck and that it did not exist before and does not have to exist now, the knowledge of how to do this is the feeble knowledge of *techne*. Unlike scientific knowledge which aims to see reality as it is in itself, and this in Greek terms is knowledge of the necessary, technical knowledge relates to the production of something which might have turned out differently and might not even have existed at all. Accordingly there is always an element of chance in the outcome even where the craftsman is highly skilled and dedicated to the task. Aristotle even peculiarly points out in distinguishing *episteme* from *techne* that the poor exercise of a craft does not imply poor *techne*, for an object might have been deliberately poorly crafted to show some feature of the craft as, for example, a potter might show beginners what happens when a piece of clay wobbles on the wheel and how best to recover the piece when this happens. Sometimes the result is poor for unpredicted reasons (such as when the electricity was cut as the cake was cooking in the oven). Because of its contingency the appearance of the artefact is dramatic just like a work of art, as Benjamin pointed out. There is an occasion to celebrate the contingency of its appearance and the historical life which enabled it to appear.

So two issues need to be brought out clearly on the basis of this reading of Aristotle. First is the link between *techne* and chance and second the link between *techne* and value. Aristotle notes that even when technicians work to the best of their ability *techne* depends on chance (*tuche*), meaning that products are essentially fragile and that they make their appearance on the back of a general fear that something might happen to prevent them occurring. While *techne* is purposeful in what it brings into existence, there is less control of the process than in the epistemic circle and the end result is not always as the producer would have wished. Poor materials, inhibiting external conditions or even inhibiting internal obstacles can mitigate an outcome and these fall under the heading of *tuche*. Aristotle quotes Agathon: 'Chance is beloved of Art, and Art of Chance' (NE vi, vi 1140a 20). The second issue is that some value is involved in every production – the production of weapons, the production of food, the production of scholarship, even if these productions are only regional features of a person's life and therefore not representative of the totality of a life. Visitors to the must-smelling dungeons of Ghent castle are shown instruments of torture all presented for the visitor to see. These artefacts not only make a statement about the ingenuity of the designers and the subtle degrees of pain they were proficient in inducing but they also make some kind of moral statement. They may have been instruments of law and order for some, instruments of sadistic delight for others, but they have a moral meaning, however lost in the mists of time it may be. Associations like these point to the fact that *techne* is an important generator of historical culture, that it reflects the values of the times and echoes the many justifications and rationalisations people used in pursuit of the 'good' life. These are always open to moral challenge on the basis of subsequent historical cultures. With epistemic knowledge, in contrast, there are no such shades. A valid syllogism as a product of reasoning remains true at every time and place no matter what one thinks about its relevance, significance or value.

The third kind of truthful thinking, the third kind of truth related to the movement of the mind is *phronesis* or practical wisdom.[1] *Phronesis* is a less visible movement and it is subject to vulnerability and history but signals a disposition to lead the good life, the *eu zen* (NE vi,v 1140a 28). Like *techne, phronesis* has to do with the contingent order and what might have been otherwise. But unlike *techne, phronesis* describes one's total disposition in life. Furthermore a disposition of this kind depends squarely on 'action' which roughly means that it summarises all one stands for in what one does. Like the productions of *techne,* the actions of *phronesis* depend on repetition building up to a habit. In this total enterprise nature indicates that we are destined to be happy, that inbuilt into our organism is a natural desire and wish to be happy. Nature also indicates what to do to be happy, for to function well as an organism one must develop one's own range of excellences; one must become educated and capable of action in the higher spheres of culture. For good or ill, however, time is needed for the development of such habits. Once achieved, even in less perfect ways, this desired state of being, this totality provides a temporary stability analogous to *episteme* but quite vulnerable to particular actions of intemperance and incontinence, in short, to vicious actions. It is a bone of contention whether this temporal residue accumulates naturally and belongs to some clear natural disposition irrespective of the agent or whether it is a contingently established residue, normally dependable but ready to reverse nevertheless, following a number of impalatable actions.

While Aristotle wants to highlight the difference between *techne* and *phronesis* by distinguishing production from action, he may have overstated the case if he meant to exclude *techne* from practical wisdom. Our view is that because reflexivity is part of the productive process of technical thinking (the craftsman in the act of crafting something wonders whether he or she is succeeding and is thus questioning his or her own skill), so a residue of a certain kind deposits itself within the psyche of the skilled practitioner and this is practical wisdom of a regional kind related to an art. Moreover in this same regional sense the one who learns to play the harp and becomes a harpist in the particular circumstances of life engages in moral action, an action of soul. In addition there is an easy link between the development of one skill and the revaluation of all one's skills, capacities and values as part of the total audit effected by *phronesis*. *Phronesis* enables the bearer to take the right course of action when pressed to do so, to step into the future with comprehensive awareness but only on the basis of the development of micro capacities which are then judged in the context of one's life in general.

So the point of this discussion concerning the first three intellectual virtues is the fact that each of them in their own way makes a claim to monumentality in thinking. The weakest claim is made by scientific (epistemic) thinking where the monuments are proposed as principles which operate as points of reference from which to set sail or to return after a mental journey of assent or dissent. In temporal terms, the mind requires some stability in order to find its way, whether this is supplied by evidence at hand or else by the principles or hypotheses in the light of which the research is undertaken. Products are clearer as monuments than scientific principles because the habits of mind of *techne* are subject to luck or chance (*tuche*) and are inserted in history and continually vulnerable to forgetfulness. While the products of *techne* make no claim to necessity, their residual quality acts as a stability to enable production to be maintained. In the case of prudence (*phronesis*) ideally what remains is a habit of mind and action affecting the whole individual and determining whether a person lives well or not. This generates the experience of being-in-time and it generates the basis for creating an environment in which it is possible to live the good life. So we have so far three movements of the mind and three ways of being-in-time – the way of *episteme,* the way of *techne* and the way of *phronesis.* It is this relation to time which gives each way of being-in-time a 'truth attaining rational quality' an *einei hexin alethe.*[2]

Where is all this leading? My basis hypothesis is that something within culture has affected monumentality and I am particularly concerned in these pages to show how this wobble has affected technology. The ability of culture to provide a time and a space for the temporal manifestations of technical artefacts is crucial to the historical meaning of technology as well as to its cultural impact. We still need to know a good violin when we see it and to know what the purpose of this object is and perhaps how to use it; we still need to be able to recognise a good performance and to appreciate the basis on which it rests. Otherwise without technical knowledge we have lost our bearings in our culture.

2. Heidegger: Techne *as Monument*

Obsolescence refers to the inability of any culture to leave its mark on history. Two senses of obsolescence compete with each other to confuse matters. If I take a dated model of car, I can look at it and consider it past, bypassed, an obsolete product. This is the first sense of the word obsolete. But if I live in a culture which knowingly or unknowingly repeats the products of the past without recognising its own time, then this means that nothing that is present can actually refer to historical time at all and we are living in an obsolescent culture. In this second sense, nothing can appear as 'obsolete' since everything is taken up, transformed, renewed, made relevant. Hence the new Volkswagen Beetle: not entirely old, not entirely new, somewhere in between. The culture of obsolescence conceals the fact that there is no present time or, more accurately, that the past has been abolished and taken up again, renewed, represented and transformed in an eternal present which transforms stable cultures into nomadic cultures. Everything in obsolescent culture exists outside historical time.

At first glance this issue seems completely disconnected from Heidegger's work on technology and we need to dig deep to find the link. In general Heidegger's account of technology in *The Question of Technology* is sometimes contrasted with Aristotle's account because Heidegger concerns himself primarily with ends rather than with means and therefore with values (Hood: 1972). As we have seen Aristotle focuses on *techne* as a skill or art delivering into view an artefact which either resides only in the soul or at least resides in products left behind in memory of the work. Sometimes Aristotle is accused of giving too much energy to means and not enough to the ends but this is understandable because *techne* provides a way for humans to create something that, were it not for the technician's skill, would not come about at all. Heidegger on the other hand sees *techne* as a way of shepherding the appearance of reality; *techne* engages in a strategy of facilitation and promotes ways of letting the essence of reality appear through a work of art. This sounds appealing but it raises once more (after Nietzsche) the problem of mysticism and it overlooks Aristotle's important claim that *techne* only mimics *physis*. Let us look at this further.

The gist of Heidegger's position is that if the artist did not create the work of art, nature itself would still manifest itself in expectation of its eventual revelation through art. The pertinence of individual, concrete human action relates to a general epiphany which originates outside man and in nature itself. Indeed Heidegger situates technology in the context of metaphysics or what he calls *meta ta physika* where *physis* refers to the 'self-forming prevailing of beings as a whole'. In remaining subordinate to *physis, techne* promises to respect and acknowledge the prevailing quality of nature and to avoid whatever endangers this. Heidegger implies that, since man is also a natural being in relation to nature as a whole, he is bound in relation to nature in himself, or human nature as part of the general unconcealedness of nature. Human intervention plays a pivotal role so that while *physis* is continually bursting forth in revelation only humans have until now found a way to give expression to this revelation; only humans have generated forms for conscious attention through paintings and other media. Without this poetic contribution, nature would remain in concealment, hidden from view and from celebration – a hidden sphere of being unrecognised, and uncelebrated. *Techne* is that which coaxes nature away from its concealment and enables it to stand in full view. Time too is critical, for revelations of this kind are fleeting and the ability of *techne* to hold them in open view varies considerably. The concept of enframing is key here.

Lovitt helpfully lists the following associations with *techne* from Heidegger's vocabulary and they are all plays on the word *Gestell* or 'frame.' *Techne* is a strategy to 'set-in-place' (*stellen*), to 'order' (*bestellen*), to 'produce' (*herstellen*), 'present' (*darstellen*), to 'represent' (*vorstellen*), to 'entrap' (*nachstellen*), to 'block' (*verstellen*) and finally to 'en-frame' (*ge-stellen*) so that the artefact is enabled to 'stand-in-reserve' (*Bestand*) (Heidegger: 1977, 15, n.14). By unfolding these forms of *stellen,* Heidegger maintains the priority of nature in the appearance itself. *Techne* and human action is subordinate to *physis* for it is nature which calls *techne* into existence. Technology as a science of means only is according to these associations a systematic implementation of means over ends. A possible distortion, a danger, is hinted at in Heidegger's text: 'Does this revealing happen somewhere beyond all human doing? No. But neither does it happen exclusively in man, or decisively through man' (Heidegger: 1977, 24). Heidegger reckons that instrumentality is a danger within technology itself for it raises once more the secret ambition of the human to rule the world by shaping the world according to human needs. Heidegger presumes the priority of *physis* over *techne* and holds that the destiny of *techne* is to bring something into presence. His text is somewhat mystical:

> The essence of technology lies in Enframing. Its holding sway belongs within destining. Since destining at any given time starts man on a way of revealing, man, thus under way, is continually approaching the brink of the possibility of pursuing and pushing forward nothing but what is revealed in ordering, and of *deriving all his standards on this basis.'* (Heidegger: 1977, 10)

Let us pass over the irony of the text I have italicised for the moment and note Heidegger's worry that indeed truth (as unconcealment) may be happening now less in the case of modern manufacturing than in the case of craftsmanship (Heidegger: 1977, 14). Let us take up the point that modern technology is not *poiesis* but something more violent, a challenging (*herausfordern*) of nature, a demand that artefacts store up energy for future use. Heidegger presents himself as a peacemaker, complaining about the violence of modern technology and arguing that it has moved far from the plough or the wood-cutter in the Black Forest who maintains the woodland as a natural resource. Modern technology on the contrary founds itself on an act of violence – the petrochemical industry, deforestation, the damming of rivers to produce electricity. In these cases where *techne* is not a *poiesis* linked to *physis* but a type of throttling of nature to deliver resources for sale to this present generation, it sets unreasonable demands on nature and in doing so subjects nature to the will and dominance of human power interests. This turn towards violence is a danger and Heidegger is led to distrust a mere anthropological definition of technology on account of it (Heidegger: 1977, 21).

What should happen is that technology should respect a granting by nature of itself in the space of appearance on the grounds that 'only what is granted endures' '*Nur das Gewährte währt*' (Heidegger: 1977, 31). Heidegger's insight centres on the human ability to 'set upon' some aspect of nature (not to throttle it) and to frame it in such a way that its very nature and its ability to endure through time is allowed to shine. He refers to the Rhine and his case could be bolstered by developing the example he gives. The great river Rhine, to use Heidegger's terms, can be framed in such as way that it never loses its link with nature. A tourist operator for instance decides that he will run a bus to a location of extraordinary beauty. He plans this in advance and secures the bus and driver to do this and he provides the tourists with a commentary which acts as a discourse to animate their experience. The three causes – final, efficient, formal – combine with the material cause which is the Rhine itself to set up the place (*gestellen*) as a standing reserve of being (*Bestand*), a place where the being of nature (*physis*) can manifest itself in wonder to the tourists. There seems to be little danger in this context of considering the river simply as a human construct. The frame, however, is human: it is a product of technical thinking, but the whole experience lies outside man's control. Nature itself remains the dominant power and a recurrent source of revelation while serious problems arise if man seeks to be the dominant force.

At first glance it may not be possible to see the shadow of the superman here but let us look more closely. As a poetic strategy, what appears is not only the thing, the artefact produced, but nature itself. Hence *techne* as poetic action does not create the artefact *ex nihilo* (as Aristotle suggested) and cannot fully control all the conditions under which it will appear but instead engages in an action of framing or enframing (*Ge-stell* p.11) which heralds the (possible) appearance of the artefact in its very being (*physis*). The lowly role of the artist here seems quite contrasted with the dominance of the superman. On this reading even the cleverest action of a craftsman cannot dominate the action of nature in showing itself, because what reveals itself in an artefact is the being of that which is represented. So the poet is the one who skilfully, craftily enables something to emerge from unconcealment into appearance which nature itself was ready to reveal. Hence it follows that the human historical contribution is significant but not originating,

> Man can indeed conceive, fashion, and carry through this or that in one way or another. But man does not have control over unconcealment itself, in which at any given time the real shows itself or withdraws (Heidegger: 1977, 18).

It is here, however, just when Heidegger's case is at its strongest that the problem of the superman as well as the problem of technology as an historical monument comes to the fore. Remember Nietzsche: the only way of setting up a monumental history is to set aside the urbanity and indeed the banality of life and to proclaim the greatness of man. Nietzsche's monumental attitude focuses on great deeds and heralds the greatness of persons who are willing to use great deeds to stand on the shoulders of past events before boldly stepping into the future (Nietzsche: 1983). This is not the attitude of the antiquarian but it is monumental because such great monumental men (usually men) etch their supposed mark on history by rising above the masses. They have in a sense become divine Dionysian men, supermen. My fear is that Heidegger too in speaking of destiny, about being called, about being the agent of unconcealment has become an eager apostle of the muted religious tone of Nietzschean promises, particularly of the promise of divinity and the illusion of greatness which he thinks the artist can achieve through proper humility. Heidegger's poetic attitude generates a mysticism around the appearance of objects whereby the operator of *techne* becomes the conduit for the greatness of nature. Should all religious people flock to his side? I think not, especially when he deflects the ethical problem of technology into a problem about ends. For Heidegger, the ethical problem is how technology can justify a clear change in ends from the tree growing in the wood as a natural event to the chair you may be sitting on as an artificial event. He does this by realigning these ends and attaching them to the same non-human origin – *physis* and in so doing it would be hard for him to see tables and chairs as anything other than deviations and abrogations from nature.

The reason I am particularly hard on Heidegger here is the fact that he has forgotten the lesson of Machiavelli. This lesson simply stated is that the lowliest of servants can be encouraged to conceive of himself as the greatest of masters in the service of something entirely beyond himself. Such a person is eminently ripe for manipulation. This is a danger also when religious practice centres on the 'otherness' of God which can all too easily lead to a Manichean expression of unquenched and 'unredeemed' appetites. This abandonment of history as a way of life was a danger too for National Socialism where the crowds could temporarily forget their own lowliness and pledge their allegiance to the absolute power of the *F_hrer*, also deifying themselves in the process. It seems reasonable to conclude that in Heidegger's account the lowly artist, servant of the great epiphany of being, is through the agency of *physis* a concealed superman, riding on the shoulders of a force that is beyond good and evil, unaccountable to human society with a permit to operate above the perceived common norm. This new ethic, however, is hardly good news.

By contrast I prefer to see *techne* as a step into the fragile reality of human culture which is a discrete and homogeneous form of action subject to its own laws and procedures. This means that *techne* forms and shapes wood, for instance, according to human ends apart altogether from the wood's inherent end to remain a tree (See Aristotle, *Physics* II.1. 193a 13-14), and in doing so *techne* is neither offering the wood or God a disservice. Tables and chairs are not poetic representations of the being of trees or indeed of being itself; they are human constructs which have a value and a justification solely within the human historical context. Let us be careful then about bringing God into it.

To deny the critical importance of these fragile elements, in my view, is to overlook the central connection between technology and the historical establishment of culture. This is why in Heidegger's account the historical monument of *techne* has been replaced by a 'standing reserve' from which point the ahistorical epiphany of nature can take place. His analysis does little to alleviate the real danger of technology allying itself too closely with power and forgetting its cultural role among the mediocre artefacts of historical culture.

Conclusion

Our subject was to describe the phenomenon of technology as monumental history and to this point the bare elements of such a description have been made. Unhappily I recognise that my attempts to see technology as history, indeed to value it as history, is being borne away by the culture which infects me and which mediates another message. Theoretically I return to the anthropological account discounted by Heidegger as too dangerous, dismissed by Nietzsche as too mediocre, and instead of endorsing the agony of greatness proposed directly by Nietzsche and indirectly by Heidegger, I prefer to measure my life against the tiny achievements and failures which, despite their mediocrity and urbanity, announce their presence as the authentic and dramatic manifestations of historical beings. Theoretically I return to the issue of monumentality and the sense of loss if *techne* is not able to furnish us with the stable products against which we can take our bearings as ordinary beings-in-the-world.

On the other hand, my culture draws me back to its own belief system which means that I have a fascination with new inventions and new products. I also secretly celebrate what is obsolete, what has already fallen behind. Whether this is microchip speed or memory capacity in the computer or pixel ratings in digicameras, I experience a certain joy in celebrating obsolescence, in having outlived these inventions and having recognised that the machine of a slightly lesser speed or capacity which I did not buy last week is in fact now old and degenerating by the minute, that my finances in relation to these purchases have remained intact despite the temptation of a week ago, that I have not been duped to invest in something that is now past and spent. I take pleasure in shaking them off like dust off my sandals. So as I take up the papers to read about what is new, I also wave goodbye to what is old and it all amounts to an experience of lording it over technical products by outliving them, of seeing the Towers come down and surviving the event, of outliving the audacity of these buildings to lord it over me as they stood there majestically pointing to the sky.

There is an irony in these feelings, of course, for having survived such monuments, I have now run out of monuments and have become obsolete myself. Consequently the joy of surviving technological change is tinged with the awareness of never being able to catch up with it and of being myself in a certain sense outmoded, past, part of the flotsam of human creativity and invention. No matter when I make a purchase now, I know that I own something that is already obsolete. My joy at being immortal has returned to haunt me and cruelly I recognise that I too have become old and washed out, unable to keep up with what is being invented, unable to do anything else but buy in to an obsolescence which I would like to avoid but which, despite the gadgetry I surround myself with, signals only that my life lacks synchrony and that it has run out of time.

The issue of technology as monumental history has therefore offered a stark choice, to opt for history or for immortality in all their many forms. Which option is human? Philosophy advances arguments for both and says that either is human. Christian theology, however, indicates that only one of these options will fulfill the destiny of historical beings.

Endnotes

1. Throughout this analysis I have used Dunne, Joseph (1993), *Back to the Rough Ground*, (Notre Dame: Notre Dame University Press). I differ from Dunne's analysis on the issue of how *techne* mimics nature, a fact that will become clearer in the next section.
2. Translations are from the Loeb edition by Rackam, H. (1999 (1926), in *Aristotle: The Nicomachean Ethics* (Cambridge, MA and London: Harvard University Press).

Bibliography

Arendt, Hannah (1977), *Between Past and Future*, (Hammonsmith: Penguin).

Barker, Andrew (1999) review of David Roochnik (1998), *Of Art and Wisdom: Plato's Understanding of Techne*, (Philadelphia: Philadelphia University Press) in *Classical Review*, 49, 2:432-3.

Benjamin, Walter (1970), *Illuminations*, 2nd edition (London: Fontana).

Goodwin, Andrew (1988), 'Sample and Hold: Pop Music in the Digital Age of Reproduction,' *Critical Quarterly* 30:3, 34-49.

Heidegger, Martin (1977), *The Question concerning Technology and Other Essays*, trans. with an Introduction by Lovitt, William, (New York: Harper).

Hood, Webster F. (1972) 'The Aristotelian versus the Heideggerian approach to the Problem of Technology' in, Mitcham, C, and Mackey, R, (eds.), *Philosophy and Technology*, (New York, London: Free Press).

Jencks, Charles (1991), *The Language of Post-modern Architecture*, (London: Academy).

Nietzsche, Friedrich (1983), 'The Uses and Abuses of History for Life' in, *Untimely Meditations*, trans. Hollingdale, R. J., (Cambridge: Cambridge University Press).

Tapscott, Don (1998), *Growing up Digital: The Rise of the Net Generation*, (New York: Mc Graw-Hill).

CHAPTER EIGHT

Nihilism or Salvation?: The Challenges of Global Technology for the Humanities

Mark Dooley

Introduction

The debate surrounding whether or not science has the upper hand over the humanities is still one that rages in our universities today. The most concrete manifestation of how strained relationships are between these two groups can be detected in the resentment displayed towards scientists when they, as per usual, receive far higher levels of funding than their counterparts in the humanities. Such justifiable resentment has, I believe, served to exacerbate the age-old inferiority complex that we humanities educators have developed when we think of ourselves in relation to the scientists. This inferiority complex is understandable when one considers how undervalued, not only in terms of financial backing but also in terms of prestige, the humanities have become. If there is a 'crisis' in the humanities some of the blame can undoubtedly be placed at the feet of those who consider the arts as little more than navel-gazing.

That acknowledged, however, many of the problems that currently beset the humanities are, I believe, self-inflicted. Speaking from my own particular perspective, philosophers and theologians never really got over the challenges that the New Science posed in and around the sixteenth century. They continued to think of themselves as the sole interpreters of the universe, as those divinely inspired few who had all the correct answers. This belief that they were purveyors of the 'Truth' led philosophers and theologians not only to demonise the new scientists, but also many of their fellow-humanists. Even today one can continue to detect in their attitude to the literary critics, the historians, etc., a certain condescension based on the view that the really serious work happens in the philosophy and theological departments. The literary critics, so the argument goes, fail to ask the ultimate questions and are thus not as qualified to speak authoritatively about matters of pivotal concern. The philosophers and theologians, that is, look down on their humanistic counterparts in much the same way as the scientists look down on them.

Factitious Distinctions

I intend to proffer some suggestions as to how these rather arbitrary and factitious distinctions between the sciences and the humanities, on the one hand, and the various branches of the humanities on the other, can be overcome in the closing stages of this article. For the moment, however, I want to dwell further on the effects that such divisions have engendered. The success of the New Science in explaining the universe, as I have suggested, led to a closing of the philosophical and theological ranks. As a consequence, philosophers and theologians went to work on a host of critiques of the scientific approach. In most cases, science was accused of being reductionistic, nihilistic, and a threat to the moral stability of the species. The philosopher, Immanuel Kant, proclaimed that in order to counter the mechanisation of the world, science ought to be considered as that which deals with the merely phenomenal, with mere appearance, while philosophy and morality were the stuff of the noumenal realm, of reality *qua* reality. This belief that science and technology are the enemies of genuine thinking, morality, and the spiritual, culminated in the last century with Martin Heidegger's polemic against science and technology in what he termed, 'the age of the world-picture' (Heidegger: 1977, 115-154). For Heidegger, we should, as far as possible, resist modern science and technology because both are nihilistic in the extreme. That is, modern science and technology represent for this philosopher the apotheosis of the slow demise of thinking, *qua* reflection on the nature of Being. Ever since we in the West invented epistemology – ever since Plato, in other words – we have been steadily and wearily embracing nothingness as if it were man's true essence. 'Nihilism,' states Heidegger, is 'the fundamental movement of the history of the West,' or 'of the peoples of the earth who have been drawn into the power realm of the modern age' (Heidegger: 1977, 62-63). It is they who have confused 'the suprasensory world, the Ideas, God, the moral law, the authority of reason, progress, the happiness of the greatest number, culture, civilisation' (Heidegger: 1977, 65), with 'the place of habitation proper to man's essence' (Heidegger: 1977, 66).

This dire diagnosis of the current nihilistic state of the West is not unique to Heidegger alone, although it is in no small measure thanks to him that it is a common view amongst many twentieth and twenty-first century intellectuals. Heidegger furnished philosophers and theologians with a stick to beat the scientists. In so doing, however, he also managed to undermine the efficacy of some of the most important achievements of the modern world. For Heidegger, as the above passages make clear, there is no real distinction between instrumental rationality and the dawn of the democratic revolutions. Both are phases in the unfolding tide of nihilism that culminates in the technological fury of the present age. Not until we have released ourselves from the enervating grip of so-called 'progress' will we have learned how to think in a manner reminiscent of the philosopher of Being. Not until then will we be 'saved'.

Affirming Postmodernism

What unites Heidegger and his disciples, as well as contemporary Marxist writers, post-colonial thinkers, a number of French poststructuralist philosophers, and many current day theologians, is the belief that modern science and technology represent the cancer that needs to be eradicated if we are to be true to ourselves. This, I have been arguing, is the most significant consequence of the hostile reaction of the philosophers and the theologians to the challenges of the New Science. What is common to all of these disparate thinkers is the view that at some juncture we took a wrong turn and we have been paying for it ever since. This assumes, of course, that there is a right way and that we have to undo what we have done in order to access it. For the Marxists, the goal of history will remain unrealised until the tyranny of capitalism has been vanquished. For some in the poststructuralist camp, the priority is to tear back the layers of ideology and subterfuge until a new being is created. For the Heideggerians, what needs to be challenged is everything that is tainted by technology. For certain contemporary theologians such as James Edwards, Nicholas Boyle, and Stanley Hauerwas, we live in an age of 'normal nihilism', one which demands that we find ways, as Hauerwas argues, to survive and confront 'capitalism, democracy, and postmodernism'. The Heideggerian temptation to conflate capitalism and democracy, the Nietzschean temptation to see the democratic West as 'the age of the last men', and the Marxist temptation to identify democracy with the lavish excesses of the Bourgeoisie, are all part of the general attempt to privilege philosophy as that which will save us from the deleterious effects of modern science and everything that has followed in its wake.

Such latter-day Marxists and Heideggerians, such as Terry Eagleton, Fredric Jameson, and Hubert Dreyfus, follow Stanley Hauerwas in branding as 'postmodernist' any form of thought that considers Heidegger's assessment of our current situation as being wrong. For Jameson, postmodernism represents the logic of 'late capitalism', while for Dreyfus the postmodern condition is one in which the internet has succeeded in making us incapable of choice and genuine commitment, thus leading to a world in which boredom rules and human lives are thoroughly unfulfilled. For the theologian, Nicholas Boyle, postmodernism is Thatcherism, while for fellow-theologian, Paul Lakeland, it amounts to fragmentation. All in all, postmodernism does not have many avid supporters.

While there is a certain truth to these assessments, it is by no means the whole truth. That is, like all movements postmodernism is not without its excesses. The work of Jean Baudrillard and Mark C. Taylor, for example, are stark illustrations of postmodernism that has lost sight of the political and ethical implications and consequences of the postmodern. There is, however, another strain of postmodern thought that is charged by a political dynamic. Postmodernism of this sort tells us that the primary motivation for any philosophical reflection is to effect political change, and that such change ought to be in the direction of greater human freedom and equality. In other words, the form of postmodern thought that I favour, the form enunciated in the work of people like Jacques Derrida and Richard Rorty, is driven by a prophetic impulse to bring to fruition the world envisaged in the Christian scriptures and in the Declaration of the Rights of Man, a world in which caste and class have disappeared and in which nothing serves to limit the right of any single individual to be understood. 'Postmodernity', on this reading, means 'after Modernity', or after secular rationality; while it applauds the political advances of the Enlightenment, it resists the view that a religious consciousness is incompatible with those advances. In fact, it goes further and suggests that the politics that we derive from Enlightenment thinking is but a working out of the fundamental tenets of the Judeo-Christian framework and tradition.

I said that postmodernism of this sort is 'prophetic', by which I mean that it is propelled not by certainty but by hope, hope for a better world to come. As such, it is inspired not by Nietzsche, whom most consider to be the father of postmodernism, but by Kierkegaard. The latter's insistence that we are ineluctably subject to time and chance and that there is no underlying 'true' self to be uncovered, but a self that is constantly putting itself into question in the name of those with whom we have not traditionally associated, are key motifs of prophetic postmodernism. Consequently, we must be suspicious of any so-called 'grand narrative' that has hitherto served to exclude or ostracise. Imagination is the faculty of faculties for Kierkegaard, for it allows us to surmount traditional boundaries so as to engender more inclusive societies. Such, for him, is the outcome of the *imitatio Christi*, or the imitation of Christ, in which the individual opens out to those whom he or she traditionally thought of as strange. So, *pace* Lakeland *et al*, postmodernism, at least as I construe it, is far from heralding fragmentation or disruption. The only threat which it signals is to traditional identities that have served to reinforce sectarian, social, and ethnic division. If anything, thus, it strives to overcome fragmentation in favour of a more inclusive and harmonious world. This is why, as I shall suggest below, it is highly compatible with the social, political, cultural, and economic effects of globalisation.

Affirming this brand of postmodernism has, however, a number of considerable consequences. In the first place, one has to reject Heidegger's prognosis concerning the West. For all scenarios that tell us that we are drowning in a quagmire of nihilism and despair are, as I suggested above, founded on the belief that man has a 'true' essence, one that he has lost sight of in the technological tumult. But, as stated, postmodernism rejects the idea of a 'real' or 'true' self. Hence, it must give up on the idea that what we have now is somehow intrinsically degenerate or out of step with the way we really are. This means in turn that all those features of contemporary life which Heidegger regards as nihilistic, and others see as ideological, are considered by postmodernists as positive steps in the right direction. In effect, therefore, postmodernists reject the denunciation of science and technology levelled by philosophers and theologians as a way of safeguarding their professional integrity. They do so because they are of the view that neither the philosopher, the theologian, nor the scientist, has all the answers, that none of them has a way of unlocking the ultimate clues to the universe, being as they are products of their time. They are as aware as Heidegger was of the dangers that science and technology present, but they are equally cognisant of the benefits that it portends. For the 'age of the world picture' is an age in which we have more parliamentary democracy than ever before, more scientific advances in health than ever before, more freedom from oppression than even our recent ancestors, and many more opportunities in so many different spheres of life. To repeat, this is not to overlook all that is wrong with contemporary society. It is simply a matter of saying that ours is a time in which there are more signs of hope that the dreams of those who sought refuge in the Christian catacombs, as well as those who plotted the democratic revolutions, can be realised.

With every age comes success and failure. When we address one set of sufferings, we inevitably give rise to yet another set. But this is nothing less than the price that must be paid for progress, and I simply disagree with Heidegger that progress is yet one more symptom of nihilism. For those of us who have no truck with Nietzsche's excoriation of democracy and the ideals of Christianity, there can be no doubting that this age, despite its obvious cruelties, has given more expression to those ideals than any we have on record. Heidegger would undoubtedly rejoin, as his disciple Hubert Dreyfus remarks, that in this age of the world picture 'everyone might simply become healthy and happy, and forget completely that they are receivers of understandings of being,' and that 'would be the darkest night of nihilism' (Dreyfus: 1993, 314). But only a philosopher who thinks that there is something more essential to bow down before would suggest that universal health and happiness are symptoms of a time of ultimate despair. For Heidegger, however, the task of thinking must take priority over the political task of ensuring that as many people start out with equal chances of education, health, and happiness as is possible. This remarkable suggestion is the result of the firmly held belief that nothing matters more than our retrieval of the forgotten question of being. Similar considerations drove Lenin, Mao, and all those who believed that fresh running water was a silly pursuit when compared with the bringing to fruition of the 'new man'.

Technology in the Service of Transcendence

Let me now turn to the question of how all of this relates to transcendence. As should be obvious from my remarks thus far, the ambition to transcend time and chance is something that postmodernists resist. They also abjure from attempting to transcend their particular age in the direction of a new self totally divorced from the one that they have inherited from tradition. While they yearn to overcome traditional differences between groups and societies, they do not believe it possible to wash oneself clean of one's language, culture, and heritage. Their so-called 'ambition of transcendence' takes the form of an attempt to blur the boundaries between peoples so as to generate greater levels of democratic and Christian fellowship. As such, they see transcendence as a matter of surmounting traditional prejudice and, in so doing, extending the same privileges as we enjoy to those who have been less lucky. Imaginative identification for the purposes of building more democratic communities is the process of transcendence that the postmodernist recommends. So where Heidegger sees nihilism, the postmodernist sees salvation. Universal health and happiness, despite the claim that we need something more than this to be fully human, is the aspiration of those who come after Modernity. Because postmodernists have given up on ideology, they no longer think that everything that the West has to offer is intrinsically corrupting. Hence, they do not look at science and technology with the same disgust as their proto-Heideggerian counterparts. Rather, they try to observe how we might deploy science and technology to advance the cause of democracy, what John Dewey called, 'the redeemed form of God'. This, of course, was not a denial of God, but just a way of saying that our democratic institutions are perhaps the greatest monuments to the Judeo-Christian strand of our tradition. That is, the West, as Richard Rorty reminds us, is not only the place to have developed nuclear weapons and the gas chamber, but is also the place where the welfare state, democracy, and the Declaration of Human Rights were developed and drawn up. In other words, the West has been quite good at coming up with ways to combat its mistakes. This is not to say that the West is superior in any way to other cultures, if by superiority one means that Westerners have more in the way of rationality or humanity. Rather, it is just to say that it has, for a whole host of reasons, been better able to utilise its natural and intellectual resources. Luck, more than anything else, has played a central role in this story. When the West used this luck to colonise and suppress, it was indeed in a nihilistic phase. But of late it has made up for its past sins by giving us all sorts of means by which we have been able to cultivate greater levels of tolerance and affirmation of difference.

Technology has, I submit, helped us in our ambition of transcendence towards a world in which universal health and happiness are the norm. This is so primarily because where technology goes democracy follows closely behind. This, of course, was not always the case, but today we have, I believe, sufficiently come to grips with the technical revolution so as to be able to control and use it to our own advantage. What certain portions of the modern West have taught us is that with wealth and opportunity goes a willingness to give democracy a chance, and in giving democracy a chance the grip of xenophobia, misogyny, and religious intolerance, is loosened. In other words, the goal, as I see it, is not, as Heidegger suggested, to destroy the West, but to render the advantages, technological and otherwise, of the West to as many as possible and as fast as possible.

Conclusion

This brings me back now to the beginning, and to the divisions between science on the one hand, and the humanities, especially philosophy and theology, on the other. I said above that I thought that for a number of historical reasons these divisions had taken hold, but that now they seem, to me at least, factitious or unimportant when looked at in the larger context. That context is one which postmodern philosophers and the phenomenon of globalisation have helped to bring about. Postmodern philosophers, as I intimated earlier, sought to help those in the humanities to overcome their inferiority complex in respect of the sciences, by urging that philosophy ought to drop the idea of a 'true' self, and follow Kierkegaard in thinking that hope and imaginative identification with those who are least like us, signal the best means by which the self can put itself into question. In so doing, they have succeeded in making the boundaries within the humanities more fluid and porous. Moreover, in undermining Heidegger's depressive account of the present age, they have made it more plausible to believe that science is not the enemy, but yet one more set of tools to help us create a 'casteless, classless, egalitarian society' (Rorty, 1999, Preface). By replacing certainty with prophesy, postmodernism helped to create an environment in which the suspicions that existed between philosophy and science dissipated.

Secondly, the phenomenon of globalisation has proved to many within the ranks of the humanities that the dreams of the French Revolution and ideals of the Sermon on the Mount might have a real chance of being made real not, as the neo-Marxists and proto-Heideggerians believed, through the demise of capitalism and Western scientific ingenuity, but through their proliferation. For globalisation has shown that the advantages to the poor, culturally, economically, and politically, far outweigh the risks. Market economies driven by Western technology have led to an interdependence never before envisioned in the history of our planet. For the first time we have a chance of bringing into effect a global government that sees health, happiness and opportunity for all as its ultimate objective. It is hard to know which comes first: peace and then prosperity, or prosperity followed by peace. Either way, the link between the two is clear and nothing can deliver both as quickly and as efficiently as the global technological revolution. Consequently, those in the humanities should now be less inclined to separate themselves off from the sciences. They should seek to have as much co-operation with them as is possible, for only in so doing will they really contribute to making their social and political dreams happen.

Now it would be foolish of me not to have some reservations about the global technological revolution. I am concerned, for example, that there is a neo-liberal conservative government in charge in the United States currently that sees in globalisation a means of making profit under the spurious guise of caring about democracy. I am concerned by the actions of the super-rich and by unscrupulous scientists who do their bidding. But, that acknowledged, we have quickly cottoned on to the dangers and are addressing them with some zeal. For example, we are no longer prepared to have unregulated markets, which means that we are seeing the value in setting up institutions like an 'economic security council' under the auspices of the United Nations. This, one hopes, would be the first step towards the demise of fully unregulated free trade. This is why it is more important than ever to get poorer countries into a currency zone, so as to protect them against the abuses of the market.

I want to close now by saying something about the role of the theologian in this global world. As with the philosopher, I don't think the theologian can survive the current climate by refusing any longer to co-operate with the sciences. Indeed, I see it as the theologian's golden opportunity to once more make a social impact. Their tendency to veer either to the left or to the right has served only to marginalise the role of theologians in our society, and that is to be regretted. By seeing in technological advancement a step towards an equal world without wholesale revolution, the theologian will free himself from both extremes and take up the middle ground. Once there, he or she will assume the role of what Rorty once again calls an 'agent of love', one who strives to put 'technocratic manipulation' 'in the service of love' (Rorty: 1991, 48). As such, he or she will stand alongside our most conscientious journalists, our social critics, and our anthropologists, in an effort to, firstly, explain how and why the phenomenon of globalisation can effect the most profound change, change that chimes gloriously with the vision of social justice proclaimed by the prophets of the Judeo-Christian tradition. But he or she will also be on hand to continually warn us of the threat to social justice, tolerance, and peaceful coexistence, of the global way of life. For example, the theologian as agent of love will bring to light the unnecessary suffering being experienced by the losers from globalisation and try to suggest ways in which we might counter such manifest inequalities. Moreover, he or she will help us to come to grips with the inevitable stream of refugees that will inevitably pour to the West from foreign lands, and will even act in advance so that our hearts are sufficiently warmed in advance of their arrival. With the benefits of globalisation come responsibilities and, in my estimation, the theologian is one of those agents who can continually arouse that sense of responsibility. He or she could do no more to keep alive the Judeo-Christian tradition in our age than this. By encouraging us to see the age of scientific technology not as one 'in which openness and freedom are rationalised out of existence,' but one 'in which the democratic community becomes the mistress, rather than the servant of technical rationality' (Rorty: 1991, 20), the theologian opts for salvation over nihilism, hope above despair, prophesy above certainty. Like my philosopher and scientist, he looks to a time in which the age of the world picture is but a propaedeutic to an age in which universal happiness and health are not regarded as insufficiently spiritual to count, but are the most honoured and cherished ideals amongst all.

Bibliography

Dreyfus, Hubert L. (1993), 'Heidegger on the Connection between Nihilism, Art, Technology, and Politics' in, Guignon, C. (ed.), *The Cambridge Companion to Heidegger*, (Cambridge: Cambridge University Press).

Heidegger, Martin (1977), *The Question Concerning Technology*, trans. Lovitt, William, (New York: Harper Colophon Books).

Rorty, Richard (1999), *Philosophy and Social Hope*, (London: Penguin Books).

Rorty, Richard (1991), *Objectivity, Relativism, and Truth*, (Cambridge: Cambridge University Press).

CHAPTER NINE

Seeking the Highest View: The e-sense of Technology[1]

John Sharry, Gary Mc Darby

There was a small village located in the centre of a large jungle. Over time, the people decided that they would like to voyage out of the forest and to make their way to the sea. To do this they consulted the wise people in the village, notably the scientist, the engineer and the philosopher. The scientist and engineer quickly took the lead in the project. The scientist made cutting tools like machetes that could cut easily through the dense undergrowth, and the engineer organised the villagers in small working teams. Over time an efficient system was evolved, the scientist made the best and sharpest tools and the engineer streamlined their application. Soon the village was making steady progress through the jungle. The philosopher, who sat in the trees all day long, seemed to make little contribution and people wondered what his value was.

One day the group came upon a particularly high tree. It stood out as the highest tree in the jungle. The workers quickly bypassed it as they efficiently made progress, but the philosopher stayed behind and climbed to the top of the tree. From the top he had a fantastic view of the whole jungle. He could see the villagers, cutting like a snake through the jungle and in the far off distance he could even see the sea. Then to his horror, he noticed that the villagers were not heading towards the sea, but in fact were moving in the opposite direction, towards a large hidden chasm. If they continued on the same course, many of the villagers would fall to their death. Alarmed, he climbed down the tree and he rushed over to the scientist and engineer leading the group.

'We are heading the wrong way, we are headed towards disaster,' he shouted.

'Shut up,' the engineer and scientist replied in unison, 'we are making great progress.'

Technology and the use of tools has characterised the evolution of humans and is one of the key differences between us and the rest of the animal kingdom. From the first humans who used axes to cut down trees for fuel or to hunt and prepare food, to modern day humans who use deep sea oil excavation to find fuel and microwave ovens to cook, people have always used tools and technology to further their goals. In recent years, the new technologies of the computer and the Internet have had a profound impact on our lives. The Internet has made the access to information almost instantaneous. If we want information about a company, we no longer have to write a letter asking for a brochure (with the frustrating delays of folding paper into an envelope and licking a stamp), we can now immediately log onto their website. If we want to consult an academic journal, we no longer have to make a physical journey to a library (and sometimes wait weeks for an inter-library loan) we can simply register with an online catalogue, e-mail the author directly or even consult the author's personal website. We no longer have to delay in hearing updates about the latest news story, as we can receive minute by minute updates on websites. We even no longer have to be at our computer to receive the news as we can arrange for regular feeds to be sent to our mobile phones or personal digital assistants. In addition, the new technologies have revolutionised our ability to communicate personally to one another. With the advent of e-mail, Internet relay and voice chat, mobile phones and text messages, we have ever more ways of staying in touch with our friends and family, and with the world wide web at our disposal we have increased exponentially the number of people we can instantaneously talk to. We can log into chat rooms and engage in conversation with like-minded people even though they live thousands of miles away.

Though there have been benefits, the impact of the new technologies has not been all positive. Though much of technology has been about providing us with labour saving devices, whether this is motor cars, microwave cookers washing machines, faster computers or whatever, ironically, technology has accelerated many aspects of our lives and put more pressure on people to do and produce more. We are now flooded with hundreds of e-mails that clamour for attention; we are instantly at the beck and call or our mobile phones and pagers; and we are driven by our ever faster computers to be more productive. As a result, in our technological age, we can find that we have less time in the day and are under more pressure, meaning that many of the most valuable things in day to day life get squeezed. We now have 'one minute bedtime and birthday stories' for very busy parents to read to their children, and many people are spending less time talking to their families.

In addition, though we have many more ways of communicating with one another (via mobile phones, e-mail etc.) and many more people to potentially communicate with (via an open Internet), it must be asked whether this has led to a real increase in human connectedness? Instead of meeting friends or sitting talking to parents, teenagers are now more likely to spend hours on the Internet relating to strangers. In families in the evenings, parents exhausted from long working hours in front of a computer screen, collapse in front of the television as a passive recipient of this 'old' technology, while their partner watches a separate TV in another room, and their teenage son or daughter taps away at a keyboard in their bedroom talking to Internet friends. Despite great claims about the benefits of the Internet, researchers have found that greater Internet usage is associated with increased social isolation and depression and less communication with family and friends (Kraut *et al.*: 1998). Hubert Dreyfus (Dreyfus: 2001) argues that the Internet deprives users of the embodied experience that is essential to meaningful human interactions. The nature of the Internet can cultivate anonymous uncommitted encounters. Rather than encouraging people to become involved in causes, the Net can encourage voyeurism or social apathy; people surf without the risk of social involvement and this can lead to a general malaise. In addition, we are unlikely to trust people that we have not met in person or make meaningful agreements with them. As Seely-Brown and Duguid argue:

> Generations of confident videophones, conferencing tools, and technologies for tele-presence are still far from capturing the essence of a firm handshake or a straight look in the eye (Seely-Brown & Duguid: 2000, 4).

Obsessed by its benefits, people are often blissfully unaware of the downside or the limitations of technology. As John Naisbitt states, we have become fascinated by technology and the almost magical things it is capable of, whether this is cloning a person or creating microscopic computing (Naisbitt: 2001). In our personal lives, we are enthralled by the newest technological gadgets, whether this be mobile phones that play music, or wrist watches that act as personal digital assistants. This fascination has become an obsession in our lives to the extent that we accept the relentless progress without question and are in danger of losing the ability to step back from it and to choose what role we really want technology to have in our lives.

e-sense: A New Sense of Technology

Media Lab Europe is a research institute concerned with the development of new technologies that enhance and expand human potential. Our vision is to be at the forefront in the innovation of 'cutting edge' technologies, but also to provide a space for critical reflection on how these technologies can contribute to the evolution of humanity. The aim is not to be 'caught up' in a race for new technology simply for its own sake, but to stimulate discussion and reflection on the broader implications of technology and its role in people's lives. Human values are central in this process and the aim is to make technology sensitive to our individual and collective needs. One of the metaphors that have evolved in discussions at Media Lab Europe, has been the conception of future technology as potentially providing us with a 'new sense' in the world. The emergence of this 'e-sense' is about providing us with access to new data and information (like a new physical sense), in a way that is sensitive to our human nature and which also helps us make sense of and interpret this data. In addition, just as the human senses are universal, we believe this 'e-sense' should be universal and accessible by all people. This would ensure technology is open to all and centred on our individual and collective goals.

The Human Senses

As a physical scientist it is easy to simply conceive of the five human senses as operating in an analogous way to physical scanners, such as X-Ray or Magnetic Resonance Imaging machines, that gather 'objective' information about the mass of moving particles and atoms that constitute the world. But this is an over-simplified view. The human senses are integrated into the human body, come with a remarkable system of perception, and are integral to human understanding and subjective meaning. Human perception provides a remarkable system to order sensory data that deletes the irrelevant and redundant, ensuring that what we perceive is based on our goals and concerns. For example, when we see the minimal data of a circle with two dots in the upper third and a curved line in the bottom third – we perceive a face – one of the first interpretations the human baby maybe makes about the world. We automatically and with ease, perceive the distance, height and the relative movement of objects without conscious effort. Our perception allows us to adjust to different contexts and inputs. Our eye automatically adjusts to changing levels and types of light to maintain a quality picture. Our ears can pre-select certain sounds that we want to hear. For example, when talking across someone at a party, our ears can tune into the person we are talking to, and tune out of the speech of another person, even though they might be physically nearer to us.

In addition, our flexible sensory perception allows us to have complex sensory experiences. At different times when listening to a piece of orchestra music, we can tune into the melody or the harmony, we can perceive the sounds of the individual instruments or concentrate on the holistic sense of all the instruments playing together. We also use our senses in combination with one another to create a meaningful understanding. For example, our sense of taste works in close conjunction with our vision. The rich sensory experience of eating an orange is created by the visual sight of the orange, the smell as it is peeled, the sound as it is squeezed as well as the resulting taste on our tongue. Finally, the human senses constitute and are integrated into meaningful human experience. This is the difference between the objective understanding of the smell of a rose as simply a certain combination of chemicals and the utterly subjective experience of the scent of rose replete with so much meaning and pleasure. Or it is the difference between seeing a stranger in a room and recognising a loved one whom we have not seen in many weeks.

The Human Senses and Technology

In many ways, technology has been successfully used to expand the range of the human senses. For example, though we can only see visible light we can wear night-vision goggles that allow us to 'see infrared' and thus be able to see at night. With the aid of X-Rays and MRI we are assisted in being able to 'see' through solids and with aid of a microscope we can examine small objects way below the distinguishing size of our unaided eyes. Similarly, with radar we can detect moving objects that are far away and with telescopes we can discern terrestrial bodies beyond our field of vision. In addition, technology allows us to sense events that are distant from us, whether this is live TV or an internet broadcast, or the 'old' technology of a telephone. We can also re-experience events that occurred at another point in time, whether this is listening to a CD of a rock concert or watching a DVD of a movie.

Despite the expansion and 'extra sensory' input that technology has afforded the human senses, there are a number of specific limitations. Firstly, technology is perceived as something distinct and not tailored to the person. Whereas the human senses are part and parcel of our bodies and move with us in the real world, our interface with technologies is generally via the tedious mechanisms of keyboards, screens and buttons, based on what suits the machine rather than what suits people. The interfaces often require a great deal of knowledge to use, can be rigid and limiting and not streamlined to the individual person like a human sense. Dreyfus argues that it is precisely the lack of appreciation of human presence and our bodily based existence that leads to many of the limitations and associated problems of the internet (Dreyfus: 2001).

Secondly, technological advances have not matched (or come close to) human perception. Indeed, rather than technology ordering information relevant to our concerns, we can have the experience of it over-whelming us with too much information, and interrupting us and distracting us from what is important. When we do a search on the internet the search machine brings us back three million references rather than the three we wanted to access. Or we are constantly at the beck and call of pagers and mobile phones vying for our attention. As Ken Haase notes, much of our experience of technology is not characterised by artificial intelligence, but rather artificial stupidity. Unlike human perception, that learns and can adapt to new contexts and situations, computers are characterised by isolation, rigidity and inflexibility. As yet, they do not appreciate subtle changes in context, and simply crash if their input changes drastically (Haase: 2002).

e-sense and Technology

The stones shall speak and the walls shall listen.

While it may be impossible or at least undesirable to create technologies that replace the human senses, it is conceivable to envisage making technologies that are more 'sense like' and that complement the intelligence of the human perception. Current innovations focus on making technology more sensitive to the fact that we are body-based creatures who move about in the real world. For example, this has led to the development of objects that we take with us, such as mobile phones ('stones that speak') and intelligent environments such as smart rooms ('walls that listen'). The traditional 'keyboard, screen and push button' interface with computers disappears and is replaced by new novel interfaces that are more 'sense-like' and thus more in tune with needs of people.

In Media Lab Europe (in particular in the MindGames team), we are interested in developing such novel 'sense-like' interfaces such as wearable computers, that resemble ordinary clothing and accessories (e.g. wrist watches, glasses, belts etc.) Such computers will contain non-intrusive sensors that can read a range of biometric signals from the body such as, temperature, heartbeat, skin conductance, brainwaves and even information about gestures and facial expressions (via sensors that track movement and cameras that can analyse facial expression). An integrated system such as this would be able to provide individuals with feedback and give them a better sense of their internal state and keep track of their health. For example, the biometric sensors could record over time a person's heart rate, temperature and blood pressure, providing an early warning system for disease and health problems. Such information would be invaluable when consulting medical doctors as they will have access to a download of their patients' biometric data over the last few days or even weeks and months. It is conceivable that such information could be communicated in real time via the internet to allow for remote diagnosis and monitoring of people's health.

This integrated system could also provide feedback about a person's mood, emotional state and attention level. People often are not aware of their own mood. For example, a person under stress could become irritable and not realise it until they act in an inappropriate way. The sensor system could provide early feedback on a person's emotional state, so as to empower them to take corrective action. For example, a sensor system could alert the driver of a car that they were lapsing into inattention or sleep and thus allow them to take a break. These sensor systems could also be used to communicate wirelessly with those of other people and thus could be used to help people get a sense of each other's emotional state. For example, people could use this 'e-sense' to gain a sense of whether their partner or children have had a good day at work or school. Or lecturers could use it to gain an early sense of the attention level of their audience and adjust their speech accordingly.

In addition to increasing the range and quality of information, the interface to these wearable computers will become more innovative, creative, and tailored to the person. One possible interface is a pair of glasses (or a minute projector unit concealed somewhere else on a person) that unobtrusively superimposes information on their field of vision. For example, a red dot could appear in peripheral vision as a person becomes overstressed. (Of course, this feedback would be tailored to individual preferences; some people might prefer an auditory cue such as a beep, or a more relaxing image, such as a cascading waterfall, to motivate them to take action.) Such interfaces have a range of possible applications. For example, most people have the experience of meeting someone they already know at a party, but struggling to remember their name. What if a wearable computer were to take a picture of the person, scan a database for a match, and then present the wearer with the name (and whatever other information was required, such as, where they previously met, their occupation, or partner's name) in an image in the wearer's field of vision, and visible only to them.

Finally, as the technology of biometric sensors advance, it is conceivable that the wearable computers will be able to make a direct connection with peoples' minds. In the MindGames team we are able to detect via brainwaves, whether someone is thinking 'left' or 'right'. Conceivably in the future, this could be used as a means of helping someone communicate directly via their thoughts with a computer. For example, someone with minimum motor control due to disability (or perhaps if they were in a coma), could communicate via this sensor system. At minimum level they could communicate, yes or no, by either thinking 'yes' or 'no'. In the future, it is conceivable that we will be able to discern more complex thoughts and communications in this way.

e-sense and Understanding

You can bring children to a computer,
but that will not make them think.

One of the dangers of the computer revolution and the explosion of 'information technology' is that human understanding and knowledge might be simply seen as a quest for information or data finding, rather than involve creative thinking, judgement and action. In this 'information view' of human knowledge, answers to human dilemmas are thought to lie at the end of a search engine, and we seek instant and easy answers to every question, becoming passive recipients rather than active participants in knowledge creation. This misses a central aspect of human knowledge, that combines both an ability to think creatively (analysis, reflection, generation of new ideas) and act creatively (constructing things, experimenting etc). Human development and knowledge is a process that takes time. It takes eighteen years of training (and often much more) to create an adult human. As yet there is no instant download or a way to short circuit this process.

Human senses are intrinsically linked to human judgement, reflection and action. Our senses have evolved to help us make sense of the world, and to make correct and meaningful decisions. If technology in general and computers specifically, are truly to become like a new sense to people, then they must provide more than simply information, and contribute to human creativity and understanding. In the Therapeutic Technologies group in Media Lab Europe, we are interested in how technology can aid our ability as humans to think and critically reflect upon our lives, and thus help to advance our individual and collective goals. Let's consider a simple example of how this happens.

Much of psychotherapy is about helping people reflect critically about their lives and to discover new personal resources with which to solve problems. Media and technology is often used in this process. For example, when working with families of young children diagnosed with serious behaviour, emotional and developmental problems, using the Parents Plus Programme (Sharry, Hampson, & Fanning: 2003), the therapist makes a video of the parents interacting with their children, and then reviews this with them. During the review the therapist focuses mainly on what is working well and the times parent and child are relating in a successful way. In this way, the parents are helped identify and reflect about their own skills and strengths, thus discovering how to best help their children. The video feedback provides a 'constructive mirror' to the parents, inviting them to reflect about and see themselves differently. This simple technology acts like a new sense, giving parents access to new and richer data (an actual video recording) that provides a new perspective (third party), and is richer in detail than memory alone.

Future innovations, will make this technology more integrated to the person and more 'intelligent' in its feedback. For example, supposing the camera was a discreet wearable one that was linked to biometric and other sensors. A person working on a particular goal (e.g. defeating depression, learning good interpersonal communication, etc.) could programme this into the system. The camera and other sensors could then be activated at certain times during the day, either automatically by the sensors (e.g. when the person's mood is higher) or consciously by the person (when a good event has happened). These 'snippets' of experience could then be recorded for later analysis and reflection. For example, a depressed person could review the times they feel an elevated mood, and analyse how this comes about, with a view to repeating such events in their lives. Or a parent could review a happy family event or an example of successful communication with a child or partner and reflect about what was going on and how this could be repeated. This integrated technology becomes a new 'e-sense' that assists reflection and thinking, providing richer and different data than memory alone. Used in this way it has the potential to help people become self-aware, to envision personal and learning goals, and to experiment with creative options for making progress.

The work of 'Still Life' is another example of technology as providing a new 'sense' to aid understanding and learning. In this game, the computer can sense a person's movement via a camera, their state of relaxation via heart-rate monitors, and their pattern of breathing via an EMG sensor. By providing rewarding feedback in images and music, the person can potentially learn how to perform a complex movement such as a Tai Chi form, gaining feedback about their relaxation, level of attention and breathing. Thus the computer becomes an 'intelligent coach' to a person learning a complex skill that requires much practice and dedication.

e-sense as the Sixth Sense

People with the mythical sixth sense are thought to have a special ability to perceive things that other people can't perceive and to be able to discern and intuit the true nature of a situation. This special ability goes beyond just having extra sensory data at hand; it is about knowing what data is essential and how to order it so the best possible meaning is discerned. This is similar to the concept of wisdom. Wisdom is the ability to simplify the vast quantity of information that is at hand. It is about getting to the heart of the matter and picking the best decision that reflects not just pressing or immediate needs but which reflect a cognisance about wider issues and contexts.

In this chapter, we have talked about 'e-sense', meaning the evolution of new technologies that are more tailored to and integrated with the human person. To be a true innovation, this 'e-sense' needs to help people make sense of the world rather than overwhelm them with information; it needs to assist self-reflection and participation, rather than increase passivity or exclusion. In addition, if this e-sense is to be a 'wise sixth sense' it needs to link people to broader and collective societal aims, to give people a sense of the big picture rather than simply the resolution of specific or individual concerns.

e-sense as a Collective Sense?

> *Unlike humans, dolphins have a sonar sense that tells them what is in front of them and where they are going. They use this sense by sending out pulses of signals and then monitoring the reflections of these signals as they return. As intelligent social creatures, dolphins belong to families and herds (called pods) and often travel in groups. As a result they detect not only the reflections of their own sonar, but also the residue of the sonar of all the other dolphins in the group (as well as the sonar of dolphins who could be miles away). This means that each individual dolphin not only gains a sense of his own direction and intention, but also a background sense of the direction and intention of the whole group* (Haase: 2002).

Ken Haase used the above example of the dolphins' sonar sense as a metaphor for how e-sense could become a collective human sense. Imagine if technology could not only help us become aware of our own individual intentions and goals, but if it could also help us be aware of the goals and intentions of others. Imagine if the new 'e-sense' was universal and accessible by all and thus could help us develop a collective awareness as people and thus be able to chart out a shared vision for the future. With such a sense, maybe we could become more sensitive and respectful of one another and be able to share in our collective knowledge, wisdom and resources. This 'sense' would be more powerful than the 'sense' given to the philosopher who climbed to the highest tree (in the story at the beginning of the article), as his view was not accessed by the other people and he could not persuade the engineer and the scientist of the value of his information and insight. Rather with a universal and collective e-sense, we could develop a collective awareness of the important issues facing humanity at the same time. It might then be possible for collective work to creatively addressing many of the difficult issues that face us on a global scale, such as world poverty, environmental destruction, disease and war, to occur.

Endnotes

1. We are grateful to colleagues in Media Lab Europe for helpful conversations in the development of this article, particularly Kathy Biddick, Ken Haase, Síle Ó Modhrain, and Carol Strohecker.

Bibliography

Dreyfus, H. L. (2001), *On the Internet*, (London: Routledge).

Gleick, J. (2000), *Faster: The acceleration of just about everything*, (London: Abacus).

Haase, K. (2002). 'Extending Ourselves Across the Disciplinary Divides That Have Long Separated the Natural from the Artificial Sciences', (Keynote presentation, e-sense Open Day, 24th October 2002: Media Lab Europe, Dublin Ireland).

Kraut, R., Patterson, M., Lundmark, S., Kiesler, T., Mukophadhyay, T., & Scherlis, W. (1998), 'Internet Paradox: A Social Technology that Reduces Social Involvement and Well-being?', *American Psychologist*, 53 (9), 1017-1031.

Naisbitt, J. (2001), *High Tech/High Touch: Technology and Our Search for Meaning*, (London: Nicholas Brealey Publishing Ltd.).

Seely Brown, J., & Duguid, P. (2000), *The social life of information*, (Boston, MA: Havard Business School Press).

Sharry, J., Hampson, G., & Fanning, M. (2003), *Parents Plus – 'The Early Years' Programme: A video-based parenting guide to promoting young children's development and to preventing and managing behaviour problems,* (Dublin: Parents Plus, c/o Mater Child Guidance Clinic, Mater Hospital, North Circular Road, Dublin 7.

Turkle, S. (1995), *Life on the Screen: Identity in the Age of the Internet,* (NY: Simon and Schuster).

CHAPTER TEN

Ask Me Another:

An Evaluation of Issues Arising From the European Values Survey in Relation to Questions Concerning Technology and Transcendence.

Michael Breen

While Heidegger was asking questions concerning technology in 1949, he could not have foreseen the level of technological development nor the pace of change that the world has witnessed in the 54 years which have passed since then. Only 34 years ago, Alvin Toffler's *Future Shock* was examining the impact of up-and-coming technologies. He stated 'future shock [is] the shattering stress and disorientation that we induce in individuals by subjecting them to too much change in too short a time' (1970). That sense of shock is one that is often forgotten in the technological debate, where the question asked is more often 'how' than 'why'.

This chapter begins by looking at public opinion around the issue of technology and scientific advance, and correlating that opinion to other lifestyle variables such as measures of happiness and religiosity. It then continues to examine the impact of technology on society from a cultural perspective, looking at the broader issues of technological change at a social systemic level, and finally it considers the public opinion data in light of the systemic change issues. Overall the thrust of the paper is to look at some of the social implications of technology rather than technology in itself, specifically in relation to the new media technologies.

Empirical Data

We begin by looking at some empirical data about people's reactions to technology in general. The European Values Study is a pan-European project which utilises an omnibus survey focusing especially on values associated with work, religion, lifestyles and other issues. Its most recent data gathering exercise was in 1999/2000, the third of its kind and the first EVS to include former soviet-bloc countries. The previous surveys were held in

1981 and 1990. Included in the questionnaire in all three surveys were a number of items related to technology, simplicity of lifestyle and scientific advances.

Table 1 shows the Irish responses to the general desirability of more emphasis on the development of technology over the three surveys. Generally speaking the data suggest a rise in the number of people seeing more emphasis on the development of technology as 'a good thing' (62.5% to 69.6%), a corresponding diminution in those seeing it as 'a bad thing' (15% to 9.2%), and a fairly static percentage of around 20% for those who 'don't mind' one way or the other.

Table 1. Percentage of Respondents in Ireland for each option in reference to 'More emphasis on the development of technology'

	1981	1990	1999/2000
A good thing	62.5%	60.9%	69.6%
A bad thing	15%	20.2%	9.2%
Do not mind	22.1%	18.5%	21.2%

Turning to table 2, we see the corresponding percentage of responses data for the same general question, but this time the specific focus is on the desirability of a simple or more natural lifestyle. Here we see a decrease in the number of people seeing such a focus as 'a good thing' (87.2% down to 83.6%) and a corresponding scale of rise in the number choosing 'don't mind' as a response (9.9% to 14.5%). The overall variation in the numbers seeing a simple or more natural lifestyle as 'a bad thing is in the region of 1% over 20 years.

Table 2. Percentage of Respondents in Ireland for each option in reference to 'A simple or more natural lifestyle'

	1981	1990	1999/2000
A good thing	87.2%	86.9%	83.6%
A bad thing	2.9%	5.7%	1.9%
Do not mind	9.9%	7.4%	14.5%

Finally, in table 3 we see the responses to a question as to whether scientific advances are deemed helpful or harmful to mankind. Here there is little change in the numbers of people seeing such advances as helpful (41.1% to 39.7%) but there are significant changes in the other two categories. The number of people stating that scientific advances 'will harm' mankind drops (29.9% to 17.4%) with a corresponding rise to the more nuanced response of 'some of each' (29% to 42.8%).

Table 3. Percentage of Respondents in Ireland for each option in response to 'In the long run, do you think the scientific advances we are making will help or harm mankind'

	1981	1990	1999/2000
Will help	41.1%	39.9%	39.7%
Will harm	29.9%	24.7%	17.4%
Some of each	29.0%	35.4%	42.8%

If we turn to the wider constituency of the other nations included in the 1999/2000 EVS data, we get a broader picture. Looking first at the issue of more emphasis on the development of technology, we see in table 4 (overfeaf) the aggregated responses for each nation. Generally speaking, nations from the former soviet-bloc countries seem to be more in favour of such an emphasis than those in the European Union. It is interesting to note that the Irish figures are very similar to those of the overall average of all nations together. Sweden, Denmark, Greece and the Netherlands have the levels of respondents seeing such emphasis as 'a bad thing', in excess of 25% in each country. Sweden, N. Ireland, Spain, Germany and the Netherlands each have in excess of 25% 'don't mind'.

In table 5 we see similar data for the various countries based on the response to the question regarding 'a simple and more natural lifestyle'. Russia stands out as the only country with less than 60% choosing 'a good thing' by way of response. Russia, Ukraine, Germany and the Netherlands all have more than 10% of respondents opting for 'a bad thing'. The numbers selecting 'don't mind' range from 22.1% (Russia) to 9.2% (Luxembourg). The Irish figures are broadly in line with the overall average, with a slight difference in the 'don't mind' category

Table 4. Percentage of Respondents in Europe (1999/2000 only) for each option in reference to 'More emphasis on the development of technology'

	GOOD	BAD	DON'T MIND
Ukraine	89.3%	3.1%	7.5%
Russia	88.2%	3.4%	8.5%
Malta	87.8%	5.8%	6.4%
Latvia	86.9%	4.0%	9.1%
Romania	85.3%	6.1%	8.6%
Iceland	84.5%	6.6%	8.9%
Bulgaria	83.9%	5.0%	11.1%
Belarus	81.6%	3.7%	14.6%
Poland	79.6%	7.1%	13.4%
Lithuania	79.5%	5.5%	15.0%
Slovakia	79.0%	8.3%	12.7%
Slovenia	79.0%	10.5%	10.5%
Croatia	78.9%	10.8%	10.3%
Czech Rep.	76.3%	8.4%	15.3%
Estonia	75.4%	7.9%	16.8%
Portugal	71.8%	9.4%	18.8%
UK	69.2%	6.4%	24.4%
Ireland	69.2%	9.6%	21.2%
Italy	64.5%	13.6%	21.9%
Luxembourg	64.1%	19.2%	16.7%
Germany	62.0%	12.4%	25.6%
Denmark	61.9%	26.7%	11.4%
N. Ireland	61.3%	10.4%	28.3%
Hungary	60.5%	20.0%	19.5%
France	58.4%	17.7%	24.0%
Belgium	56.8%	19.9%	23.4%
Austria	56.0%	19.6%	24.4%
Finland	55.3%	24.4%	20.2%
Spain	54.8%	19.4%	25.8%
Greece	51.9%	26.2%	21.9%
Netherlands	48.9%	26.0%	25.1%
Sweden	35.4%	30.5%	34.1%
Total	69.7%	12.7%	17.6%

Table 5. Percentage of Respondents in Europe (1999/2000 only) for each option in reference to 'a simple and natural lifestyle'

	GOOD	BAD	DON'T MIND
Croatia	96.9%	.3%	2.8%
Malta	96.1%	2.4%	1.5%
Greece	94.0%	1.3%	4.7%
France	93.1%	1.0%	5.9%
Slovenia	91.6%	3.6%	4.9%
Lithuania	91.4%	3.7%	4.9%
Spain	89.8%	3.8%	6.4%
Italy	89.0%	.9%	10.2%
Slovakia	88.0%	6.7%	5.4%
Belgium	86.8%	2.5%	10.7%
Estonia	86.0%	4.1%	10.0%
Poland	85.8%	3.2%	10.9%
Luxembourg	85.8%	5.1%	9.2%
Hungary	84.4%	3.5%	12.1%
Romania	84.0%	5.2%	10.8%
Ireland	83.6%	1.9%	14.5%
Sweden	83.4%	3.0%	13.7%
Austria	82.5%	5.4%	12.0%
Czech Rep.	82.4%	6.6%	11.1%
Latvia	81.5%	6.8%	11.8%
Denmark	81.1%	3.6%	15.3%
Bulgaria	80.9%	6.7%	12.4%
Portugal	80.8%	4.3%	14.9%
Belarus	79.6%	6.9%	13.5%
Finland	79.1%	5.1%	15.8%
Iceland	77.9%	4.6%	17.6%
N. Ireland	75.9%	4.6%	19.5%
UK	74.7%	4.4%	20.9%
Germany	69.3%	14.5%	16.2%
Netherlands	66.4%	11.9%	21.6%
Ukraine	65.0%	15.5%	19.5%
Russia	59.0%	18.9%	22.1%
Total	82.3%	5.7%	12.0%

Table 6 shows similar international data for responses to the question as to whether scientific advances will help or harm mankind. 'Will help' ranges from 66.3% to 30.7%, 'will harm' from 8.7% to 18.4%, and 'some of each' from 25% to 60.7%. It is interesting to note that there is a very strong indirect linear relationship between 'will help' and 'some of each'. Ireland once again is very close to the international average point.

Table 6. Percentage of Respondents in Europe (1999/2000 only) for each option in response to 'In the long run, do you think the scientific advances we are making will help or harm mankind'

	WILL HELP	WILL HARM	SOME OF EACH
UK	40.3%	18.4%	41.3%
Iceland	66.3%	8.7%	25.0%
Lithuania	65.5%	6.0%	28.5%
Belarus	55.7%	13.4%	31.0%
Germany	49.8%	8.5%	41.7%
Sweden	44.9%	14.3%	40.8%
Ireland	40.7%	17.7%	41.6%
N. Ireland	38.4%	17.7%	43.9%
Austria	36.0%	15.7%	48.3%
Slovenia	34.5%	10.8%	54.7%
Croatia	33.9%	14.0%	52.1%
Italy	30.7%	8.6%	60.7%
Total		43.6%	12.3%

Crosstabulation analyses on the variables indicated weak but statistically significant relationships. Reliability analyses yielded non-significant results, indicating that each of these variables is measuring a different phenomenon.

So much for empirical data. But what exactly do these figures tell us? Essentially, the overall thrust of these figures is straightforward: there are considerable differences within and between countries on the desirability of emphases on technology and simplicity of lifestyle, and the overall effect of scientific advances. The picture is not by any means black-and-white. People are nuanced in their responses, particularly in relation to the issue of scientific advances. The question of the desirability of a simple and natural lifestyle is more marked across nations than it is within. Generally speaking, such a lifestyle is deemed desirable by a majority is every country but almost 20% in Russia deem that changing to such a lifestyle in the future to be 'a bad thing'.

Between nations, the greatest percentage spread, 53.9%, occurs in relation to the emphasis on the development of technology; 89.3% in Ukraine see it as 'a good thing' compared to 35.4% in Sweden. On the issue of a simple and natural lifestyle, the spread is only 37.9%; 96.9% in Croatia see is as 'a good thing' compared to 59% in Russia. In respect of scientific advances, the highest spread is 35.7% but it should be borne in mind that the number of countries involved is significantly less; 25% of respondents in Iceland see scientific advances as both helpful and harmful compared to 60.7% in Italy.

Happiness and Belief in God

It may be useful, therefore, to turn again to the EVS data, in a view to looking at some further data on a wider number of phenomena. As indicated in the opening of this chapter, the EVS contains many variables related to lifestyle and religion amongst other things. If we look at the responses to the question about happiness, we see the data in table 7. It is important to bear in mind that the lower the mean score, the higher the level of happiness. Here we see that there is a distinct break between eastern and western nations in Europe, with the western nations indicating higher levels of mean happiness that the eastern nations.

Running a one-way ANOVA on mean happiness with the attitudes to technology as factors yields a significant result ($F=55.27$, $p<.001$). The descriptive data are given along with the Bonferroni data in table 8. The data indicate that those who see an emphasis on development of technology as good are least happy and those who see such development as bad are the most happy.

Table 7 Mean Levels of Happiness from the EVS data by country (1=very happy, 4=Not at all happy)

Country	Mean	N	Std. Dev.
Iceland	1.56	965	.56
Netherlands	1.60	1002	.60
Denmark	1.61	1017	.60
N.Ireland	1.61	984	.63
Ireland	1.62	1008	.58
Belgium	1.69	1894	.68
Sweden	1.71	1012	.63
Luxembourg	1.72	1201	.59
Austria	1.74	1507	.65
France	1.76	1607	.62
Malta	1.85	1002	.70
Finland	1.86	1032	.60
Spain	1.94	1172	.59
Portugal	2.00	995	.61
Germany	2.03	1995	.67
Italy	2.05	1975	.69
Czech Rep.	2.05	1900	.54
Croatia	2.06	992	.61
Greece	2.09	1098	.72
Slovenia	2.09	979	.66
Poland	2.15	1075	.72
Hungary	2.16	991	.78
Lithuania	2.21	809	.56
Slovakia	2.26	1304	.67
Estonia	2.29	964	.65
Belarus	2.31	903	.65
Latvia	2.39	986	.68
Bulgaria	2.56	978	.81
Russia	2.57	2431	.76
Ukraine	2.57	1145	.75
Romania	2.61	1127	.74
Total	2.03	38050	.73

Table 8 Analysis of variances data from Attitudes to Technology by mean level of (un)happiness

Descriptives happiness	Multiple Comparisons Dependent Variable: happiness Bonferroni
N Mean	Mean Difference (I-J) Sig.
	(I) more emphasis on technology (J) more emphasis on technology
good 24815 2.04	
	good bad .11* .000
bad 4599 1.93	
don't mind 6207 1.98	don't mind 6.45E-02* .000
Total 35621 2.02	bad good -.11* .000
	don't mind -4.53E-02* .004
	don't mind good -6.45E-02* .000
	bad 4.53E-02* .004
	* The mean difference is significant at the .05 level.

A similar one-way ANOVA test on mean happiness with the attitudes to a simple and natural lifestyle as factors also yields a significant result (F=29.34, p<.001). The descriptive data are given along with the Bonferroni data in table 9. The data indicate that those who see an emphasis on a simple and natural lifestyle as 'good' or 'don't mind' are happier than those who see such an emphasis as 'bad'.

A further one-way ANOVA test on mean happiness with the attitudes to scientific advances as factors also yields a significant result (F=13.10, p<.001). The descriptive data are given along with the Bonferroni data in table 10. The data indicate all three groups are different on the happiness scale, those who see scientific advances as helpful the happiest, and those who see scientific advances as simultaneously harmful and helpful as the unhappiest.

Running the same set of tests on mean responses to the 'importance of God in one's life' yields some interesting results. The outcome for an ANOVA on importance of God with attitudes to emphasis on the development of technology was non-significant, i.e. there were no statistically significant differences between the groups. Running ANOVA for importance of God with attitudes to simple and natural lifestyle (F=259, p.<000) and scientific advances (F=32.66, p.<000) yields the results seen in tables 13 and 14. Those with the highest mean sense of God as important in their lives see a simple and natural lifestyle as 'good' and scientific advances as 'harmful', whereas those with the lowest mean sense of God as important in their lives 'don't mind' about a simple and natural lifestyle but believe that scientific advances will 'help mankind'.

These various analyses of variance are simply a statistical way of looking at groups – in our case groups determined on the basis of their attitudes to technology, simplicity of lifestyle, and scientific advances – and seeing where they stand in relation to other variables such as happiness, and importance of God in their lives. While it is important not to derive simplistic conclusions from the statistical analyses, it is clear that the broad trends indicate significant differences between the groupings on the variables measured. But these trends mask underlying complexities. Further analysis, outside of the scope of this chapter, should usefully focus on gender and age, as well as individual countries and varying GNP.

Table 9 Analysis of variances data from Attitudes to Simple & Natural Lifestyle by mean level of (un)happiness

Descriptives happiness	Multiple Comparisons: Dependent Variable: happiness Bonferroni
N Mean	Mean Difference (I-J) Sig.
	(I) towards natural lifestyle (J) towards natural lifestyle
good 29497 2.00	good bad -.13* .000
bad 2044 2.13	
don't mind 4217 2.00	don't mind 3.40E-04 1.000
Total 35758 2.01	bad good .13* .000
	don't mind .13* .000
	don't mind good -3.40E-04 1.000
	bad -.13* .000
	* The mean difference is significant at the .05 level.

Table 10 Analysis of variances data from Attitudes to Scientific Advances by mean level of (un)happiness

Descriptives
happiness

	N	Mean
will help	5475	1.87
will harm	1487	1.92
some of each	5524	1.94
Total	12486	1.91

Multiple Comparisons: Dependent Variable: happiness
Bonferroni

(I) scientific advances help mankind	(J) scientific advances help mankind	Mean Difference (I-J)	Sig.
will help	will harm	-5.16E-02*	.027
will help	some of each	-6.47E-02*	.000
will harm	will help	5.16E-02*	.027
will harm	some of each	-1.30E-02*	1.000
some of each	will help	6.47E-02*	.000
some of each	will harm	1.30E-02*	1.000

* The mean difference is significant at the .05 level.

Table 11 Analysis of variances data from Attitudes to Natural & Simple Lifestyle by mean importance of God in Life

Descriptives
Importance of God in life

	N	Mean
Good	30100	6.10
Bad	2069	5.20
don't mind	4352	5.05
Total	36521	5.92

Multiple Comparisons: Dependent Variable: importance of God in life
Bonferroni

(I) towards natural lifestyle	(J) towards natural lifestyle	Mean Difference (I-J)	Sig.
good	Bad	.90*	.000
	don't mind	1.06*	.000
bad	Good	-.90*	.000
	don't mind	.15	.225
don't mind	Good	-1.06*	.000
	Bad	-.15	.225

* The mean difference is significant at the .05 level.

Table 12 Analysis of variances data from Attitudes to Scientific Advances by mean importance of God in Life

Descriptives
Importance of God in life

	N	Mean
will help	5854	5.83
will harm	1667	6.40
some of each	5901	6.22
Total	13422	6.07

Multiple Comparisons
Dependent Variable: importance of God in life
Bonferroni

(I) scientific advances help mankind	(J) scientific advances help mankind	Mean Difference (I-J)	Sig.
will help	will harm	-.56*	.000
	some of each	-.39*	.000
will harm	will help	.56*	.000
	some of each	.17	.148
some of each	will help	.39*	.000
	will harm	-.17	.148

* The mean difference is significant at the .05 level.

The Impact of Technology

Survey questions, however, are a blunt tool at best for examining such critical topics in society, especially where there is such a variation in responses. To suggest that the various attitudes to technology can effectively be garnered along three axes – good, bad and don't mind – may be utilitarian and economic but it does little to advance a deeper understanding of the underlying systemic issues. To focus on partial specifics can result in a failure to direct attention to critical dimensions of the whole.

Neil Postman argues this well in his critique of technology in contemporary society , *Technopoly*. He points out that

> technological change is neither additive nor subtractive. It is ecological ... One significant change generates total change. If you remove the caterpillars from a given habitat, you are not left with the same environment minus caterpillars: you have a new environment and you have reconstituted the conditions of survival ... (1993, p 18).

New technologies have had considerable impact on social and cultural life. The influence of printing, the discovery and application of electricity, the development of mass transport, the invention of birth control drugs, the patenting of gene technologies, and the implementation of new media in society all have profound and far-reaching effects. When new technologies are implemented, they take time before they are fully adopted, but at the outset, the impact of adoption is not fully realised. It is only when a technology is fully embedded in society that its impact begins to become clearer, but by then there is no mechanism for getting the genie back into the bottle. As Postman puts it:

> New technologies alter the structure of our interests, the things we think *about* (Postman's emphasis). They alter the character off our symbols: the things we think with. And they alter the nature of community: the arena in which thoughts develop (1993, p. 20).

Postman certainly has a case to make, and his book is a well crafted response to the issue of technological domination in society today. But he overstates the case. The same problem arises with Bernard Cohen's famous comment on the mass media, that while they may not be successful in telling of what to think, but they are stunningly successful in telling us what to think about (Cohen, 1969). This is the agenda setting hypothesis that enjoys high currency in communications studies courses and one that does have strong empirical support. It does, however, suffer from the same flaw as Postman's 'technopoly' hypothesis in that it fails to take the end user into account.

The Role of Common Sense

We wring our hands, for example, at the crass excesses of the tabloid newspapers, but we also know that most readers are well capable of reading between the lines and are not as gullible as many would have us believe. People tend to exercise common sense about issues in their lives and are not simply slaves to cultural change. This is in direct contradiction to the kind of image presented to us by cult films like *The Matrix* and its sequels. The same is true of issues and concerns related to technological uptake. Most parents, for example, are concerned about the amount of time their children spent in front of television rather than studying their schoolbooks or engaging in sport and seek to limit the one in favour of the others.

We need to recognise the difference between hype and substance, especially in terms of selective media reporting on aspects of technology. The frenzied reportage about the dangers of the internet to children fails to take into account the role of adult supervision and the inherent unlikelihood of stumbling across child pornography. Interestingly enough, we do not experience the same frenzy about children being killed on our roads, or children dying of malnutrition across the world, or children being exiled to permanent poverty because of inequalities in education, although the numbers involved in these latter three categories vastly outweigh the number of children in danger from the internet. Most people, thankfully, have the wisdom and experience to make such distinctions and are active users rather than passive recipients.

Nonetheless, despite these reservations about Postman's emphasis he does have a valid point to make. Technology is all pervasive and ongoing, scientific advances continue to open new Pandora's boxes on a regular basis, and the possibility of retreating to a more natural and simple lifestyle is increasingly remote, except paradoxically for those who can afford to do it, often with the assistance of the latest technological wizardry. His central thesis is that we are collectively unaware of what is happening. Elsewhere in this volume readers will have encountered the apocryphal story of the villagers being guided in a new venture by a philosopher, an engineer and a scientist (cf p 114). Who or what shall be our guide? Postman's argument is essentially about the blind leading the blind, all caught up in a technological euphoria that does not allow for the drawing of breath nor pausing for thought.

It is precisely this failure to think critically about the issues raised by technological change that constitutes his primary cause for concern. Todd Gitlin, writing on the power of media in the twenty-first century, puts it succinctly:

> I propose that we stop – and imagine the whole phenomenon freshly, taking the media seriously, not as a cornucopia of wondrous gadgets or a collection of social problems, but as a central condition of an entire way of life. Perhaps if we step away from the ripples of the moment, the week or the season, and contemplate the torrent in its entirety, we will know what we want to do about it besides change channels (2002, p 210).

This concern is shared by theorists like Paul Virilio, who argues for an apocalypse-termination following a journey along a chronological axis as the inevitable outcome of technological development (1997). Part of Virilio's thesis is that any new technology is always self-poisoned at its inception, containing inevitably the seed and source of its own destruction, such as development of dynamite leading to the bomb or of the train to the train crash; there cannot be one without the other. Similarly, critical theorist Helena Sheehan comes to the conclusion that 'our technology has outstripped our wisdom' (1987, p 66) in the context of television content being driven by technological capability rather than a desire to communicate or tell a story. These are in stark contrast to earlier theorists like Marshall McLuhan (1964) who saw only the positive benefits to be provided by new communications media in developing a global village.

Conclusion

What we face here, then, is a deeply polarised view of technology, one which sees it as a destructive anti-social force which is inevitably damaging, and the other that sees technology as essentially benign, a boon to the world despite whatever the collateral damage along the way might be. Both are ultimately fatalistic, accepting that technology and its concomitant forces of ongoing change are here to stay, and we can do little to alter that.

The EVS data, on the other hand, do not support such theses at first glance. If the forces of technology are as irresistibly powerful as some seem to think, the evidence for a single common mindset about technological development and scientific advancement across Europe does not exist. Opinions about technology and science would appear to be quite varied and never monolithic. While there are modest associations between levels of happiness and religiosity with attitudes to science and technology, the modest nature of such associations, taken in tandem with the size of the dataset under consideration, does not allow us to draw neat conclusions about a relationship between the one and the other.

In reality, people will continue to adopt new technologies and they will also adapt to the effects of emerging technologies in society. Most theorists fail utterly to take audiences and end users into account. Despite the nay-sayers, from the Luddites onward, history shows us that society continues to grow and thrive. The real challenge to us in the face of new and emerging technologies is one that has been with us since the dawn of time: how can we, in fact, make this world a better place to live for all the people of the planet? The real risk to us is not technology; it is, rather, the possibility that some would allow themselves to be so cocooned and insulated from reality that they would not see the plight of those on the other side. The Dives and Lazarus of our time are separated by more than fine linen and good food.

The greatest challenge to society today is simply to think. Many of our endeavours are geared towards various ends, sometimes with little reflection. We can readily behave like lemmings, following on the example of others for no reason other than the behaviour of the other. Habermas's public sphere (1962, 1991, 1996), arguing for constructive open debate about the core issues by all the members of society, remains something of a dream. Technological comfort can breed ignorance as to the plight of how the other half actually lives. Despite the shrinking of our world by means of technology, we have forged a whole series of individual little worlds, rather than a single open communicating world of equals, a globe of villages rather than a global village. Insofar as technology directs us towards an unthinking acceptance of the status quo, society must be both vigilant and resistant.

An unthinking acceptance of technological change and development (and equally an unthinking rejection of such change and development) does not necessarily move us along as a society. Rather than being beguiled by the detail of the smaller picture we need, individually and collectively, to look at the larger canvas. As people we need to dream the dream of what society can be, not in terms of its technological perfection, but in terms of human freedom and fulfilment. Insofar as technology brings us toward that end, it should be embraced. Insofar as it does not, we need to ensure that our embrace of technology is not such as would exclude us from pursuing the core goals of freedom and possibility.

References

Cohen, B. C., (1963) *The press and foreign policy,* (Princeton, N.J: Princeton University Press).
Gitlin, T., (2002) *Media unlimited: How the torrent of images and sounds overwhelms our lives,* (New York: Henry Holt & Co).
Habermas, J. (1998) *The Inclusion of the Other,* (Cambridge: MIT Press).
Habermas, J. (1993) *Justification and Application,* (Cambridge: MIT Press).
Habermas, J. (1989) *The structural transformation of the public sphere: An inquiry into a category of bourgeois society,* (Cambridge: MIT Press).
McLuhan, M. (1964) *Understanding Media: The Extensions of Man,* (New York: Routledge).
Postman, N. (1993) *Technopoly : the surrender of culture to technology,* (New York: Vintage Books).
Sheehan, H. (1987) *Irish television drama: A society and its stories,* (Dublin : Radio Telefís Éireann).
Toffler, A. (1970) *Future Shock,* (London: Bodley Head).
Virilio, P. (1997) *Open Sky,* (London: Verso).

CHAPTER ELEVEN

Local Sentiment and Sense of Place in a New Suburban Community

Mary P. Corcoran, Jane Gray, Michel Peillon

Cities, neighbourhoods and communities are changing. New modes of urban living are taking shape such as the reclaiming and reshaping of inner-city neighbourhoods, the spreading of suburban estates, and the increasing concentration of commuters in peripheral towns. The suburb has emerged as the dominant urban form in Ireland over the last half-century. Indeed, it can be argued that Ireland is becoming increasingly ex-urbanised, as many of these new forms of suburban living appear to be both post-rural and post-urban. They are post-rural in the sense that vast housing estates, shopping malls and leisure complexes are colonising more and more of the countryside, threatening the sustainability of a 'rural landscape'. They are post-urban in the sense that the re-location of work, consumption and leisure facilities to the edge of the city and indeed into small towns, re-orients suburbanites away from the metropolitan core.

We know little about life in such emergent suburban forms, and less about how those who live in these places organise their individual, familial, and civic lives. Recent work in the United States suggests that social connectedness – social capital – has measurable value. Citizens in high social capital communities are more likely to have high levels of civic engagement and to enjoy a better quality of life (Putnam, 2000). Decline in the connections amongst people is associated with a decline in trust – an essential element for social co-operation. Putnam has argued that American society has experienced a decline in social capital, manifest in declining electoral participation, less volunteer activity, and less engagement in community and neighbourhood activities. Putnam's work has been challenged on a number of epistemological and methodological grounds (see for example, Edwards and Foley: 2001). In addition, Ladd (1999) has advanced a counter-argument that the stock of social capital in the

United States has not declined but rather has changed to correspond with changing social structure and political arrangements. On a visit to Ireland in 2002, Professor Putnam sounded warning bells that Irish society, which has undergone dramatic change in recent years, may be facing the depletion of its social capital resources.

This paper draws on an ongoing research project[1] a key goal of which is to identify and investigate those networks that link and bond people in Ireland's growing suburbs. Our focus is on residential locales which while spatially connected to urban centres are no longer tied to those urban centres because people who live in them do not necessarily work, seek services or socialise in the urban core. We aim to ascertain whether or not these dispersed communities are generating sufficient levels of social capital to be sustainable into the future. Here we reflect on a number of pertinent themes which have emerged from our data. Specifically we focus on how people perceive the place where they live, and how a sense of place attachment is constructed and expressed in everyday life. It is our contention that the quality and nature of urban life depends in part on the way residents relate to their urban or suburban environment (Gans: 1962, 1982; Fischer: 1982; Hummon: 1992; Varenne: 1993; Bonner: 1997). This can be looked at from the point of view of residents' identification with their locality and usage of local facilities and resources. As Tovey has argued, 'uneven spatial development in late capitalism heightens the significance of location as a source of identity and as a basis for collective mobilisation' (Tovey: 2002: 182). The ways suburban residents identify with the locality and make use of it determine in a fairly direct way the social relationships that develop there. Recent research on European neighbourhoods suggests that there is a tension between place as material reality and place as the locus of sentiment (Corcoran 2002). For example, even if an inner-city neighbourhood is degraded, residents may still feel strongly attached because they remember how the place was, and imagine how it might be. This creates layers of ambiguity in the ways people relate to the places in which they live. We need to think about the ambiguities of place as they are manifested in the emergent suburban communities, because how people feel about where they live has consequences for their capacity to form sustainable communities.

A key feature of the Irish spatial pattern has been the diffusion of the metropolitan areas into the surrounding countryside. Following a similar pattern to that which has been observed in the metropolitan areas of both the United States and parts of Europe, the boundaries of cities like Dublin have continued to expand in a seemingly relentless process of ex-urbanisation, whereby housing, shopping malls, entertainment facilities and industrial parks are increasingly located beyond the city's perimeter generally along strategic nodal points on the major road networks (Corcoran: 2000). These flows are re-arranging the distribution of population across metropolitan areas and changing the densities of both central and outlying communities (Clarke: 2003, 148).

Ratoath, Co Meath, a very small village until the late 1980s, has multiplied quite dramatically since then.[2] Ratoath recorded the highest percentage population increase in the commuter belt of Leinster in the last six years. Between 1996 and 2002, the population grew by 82% to 5, 585, (CSO 2002). Significant development in the form of semi-detached and detached housing estates has occurred around this small village, located about 30 km north west of Dublin city centre. There are approximately 2,000 young families living in the locality and more than 700 children attending the local national school. The methodological approach employed includes a standardised questionnaire survey, focus groups and in-depth interviews. The analysis presented below is derived principally from our survey findings, our observations in the field, and from our analysis of focus groups conducted with sixth class pupils in the local school, short essays on 'The place where I live' written by the same pupils, and discussions conducted with the Active Retirement Group and the Mother and Toddler groups at the St Oliver Plunkett Community Centre, Ratoath.

Choosing Rurality and Disavowing the Urban

> I really love living in Ratoath because it's really full of wildlife. The trees are growing a lot around the place. (Sixth class pupil, Ratoath NS)

In his study of Canadian ex-urbanites, Bonner debates whether or not the preference for parenting in a rural setting is the product of an anti-urban ideology or whether it is a reflection of class and occupation and not the size of a settlement. Bonner quotes Pahl, who argues that:

> this mobility through the countryside can be seen as an urban pattern – for the essence of the city, to a true urbanite, is choice. The true citizen is the one who can and does exercise choice, and only the middle and upper class minority has the means and opportunity to choose: thus while the middle class extends itself from the city into the region then, in this respect, the city has extended itself into the metropolitan region (Pahl: 1968, 271-3).

Because the city is being defined as 'choice', choosing a rural setting has to be seen, according to Pahl, as essentially an urban phenomenon. Rurality and rural culture as an object that these people claim to be choosing, cannot, by definition, exist (Bonner: 1997: 53). In Ratoath, there is clear evidence of people actively choosing to live in a place that is, to them, identifiably rural rather than urban. The majority of respondents (just under 58 per cent) moved to Ratoath from the Dublin metropolitan area. A further 15 per cent had migrated from Co Dublin, Kildare or Wicklow, with just 13 per cent hailing from elsewhere in Meath. In some respects, Ratoath, formerly a sleepy rural village, is creating a new template for modern living. Locals are in the minority, while the population profile represents a cross section of migrants from the city, towns and rural outposts in adjacent counties. It is noteworthy that the growth of suburban areas in major cities of the United States has been shown to come from the de-population from the centre as well as new growth from other regions (Clarke: 2003, 147). Our analysis of the demographic profile of Ratoath indicates that new population flows are taking people out of built up urban areas and into proximate rural localities, whereas historically the population flows were primarily from rural areas to the central cities. Historically, rural migrants to the city were poor and uneducated whereas those forsaking the city for Ratoath are largely drawn from 'an expanding middle class whose dependence on the countryside as a place to service its consumption practices is increasing' (2002, 180).

Consumer tastes are an important element of the nature and evolution of new urban areas in the countryside. There are very specific features with which people identify when thinking about and choosing Ratoath as a place to live. The most commonly cited features that attracted people to Ratoath include the village character, country feel, friendliness and sense of community. Ratoath's greenery – the hedges, fields and trees – act as important signifiers of the countryside and rurality for respondents, both young and old. Children remarked that despite all the development, and the fact a lot of people don't know each other, Ratoath has not lost 'its country look'. A participant in the Mother and Toddler group remarked that she liked the 'ruralness of the place' especially the green fields and the fact that it was 'not all concrete'. Both comments suggest that it is the aesthetic of rurality to which people are attached, rather than to the countryside itself. For the reality is that the countryside around Ratoath is under threat from development and that the very reasons that attracted people in the first place, are fast disappearing.

> I think Ratoath is a really nice place to live in because there is beautiful countryside, but there is a new shopping centre and it is going to get a lot busier, and there are too much houses which are taking up too much space. Because where all the estates are there used to be beautiful countryside and beautiful trees and animals but they are all gone. (Sixth class pupil, Ratoath NS)

The quaint country lanes which lead in and out of the village are choked with traffic taking short cuts from Navan to the M50. It is difficult to cycle or walk safely between the housing estates and the village. People frequently opt to make such journeys in their cars. As the threat posed by further development persists, the aesthetic of rurality – expressed in increasingly symbolic ways – becomes all the more important. In describing the place where they live, local children utilise a rural/urban continuum, positively evaluating the former and negatively evaluating the latter. In short, the respondents do not want Ratoath to be 'Dublin'. Not surprisingly, the children growing up in Ratoath had some difficulty in responding to questions regarding their primary identity. While most identify with the adopted county of Meath, they are frequently viewed as Dubliners by those whose families have a history in the county, and as 'culchies' by their family and friends who still reside in the Dublin metropolitan area.

Adult focus group participants were adamant that Ratoath did not form part of Dublin city, and the majority had little orientation toward the downtown. Significantly, more than three quarters of respondents in the study felt either strongly attached or attached to Ratoath.

> Ratoath is growing nearly too fast. It's a friendly town with a huge population. There used to be green fields around Ratoath; not any more – it's just filled up with estates. I would like to see the building of estates stop before we join up with other towns. Take for example Malahide: it used to be like Ratoath except much bigger but estates and buildings flowed in and now Malahide and a nearby town called Swords are joined together. I don't want to see that happen to Ratoath. (Sixth class pupil, Ratoath NS)

The spatial patterning of people's social lives (as opposed to work lives) explicitly disavows the metropolitan downtown. People express very little orientation to the city, rather operating within the triangle of Ratoath–Blanchardstown and Ashbourne/ Navan/Dunshauglin. Of those who attend church, the overwhelming majority do so locally. One quarter of respondents, however, do not attend church. This figure provides further evidence of the decline in regular church-going among the Irish population, which has been evident since the early 1990s. The majority of respondents use the shop, pub and restaurant facilities within Ratoath. Just over 40 per cent do their supermarket shopping in Blanchardstown, while more than two-thirds go to the cinema there. While more than two-thirds attend doctors' surgeries in Ratoath, the remainder travel to Ashbourne, Dublin and elsewhere for such services. This probably reflects the relatively high numbers that have moved into the area recently, and have yet to establish a relationship with a local GP. About half of respondents bank in Ashbourne, Dunshaughlin or Blanchardstown, and a further 24 per cent bank in Dublin.

Of the seventy children who attended focus groups, not one expressed an interest in visiting any downtown city facility when describing their perfect day out. Rather, they wanted to visit adventure centres, shopping malls, skateboard and motor tracks all of which facilities are located on the perimeter of the city. So a spatial pattern emerges of commuting between different suburban localities in order to fulfil everyday social, personal and recreational needs. In the process, the city and the facilities it offers are circumvented. This exacerbates the polarisation and fragmentation of the urban community already underway as a result of the proliferation of shopping malls on the city's outskirts (Corcoran: 2000).

The Urbanisation of the Rural

> I think the committee should stop giving permission to the builders to build more houses because it is ruining the place and some of the animals are dying because of it. It was fine when I came here in 1999 [but] now Ratoath is getting ruined with the new houses. The shops won't be able to handle more people coming in every day of the week. I hope the people will listen and stop building more houses in fields where some kids play. (Sixth class pupil, Ratoath NS).

Just under half of the respondents (46 per cent) expressed the view that Ratoath is a place that is changing, and as such is in danger of losing its character and tradition. While respondents generally appear to be fairly happy about where they live, a number of key problems were identified: 27 per cent of respondents expressed concern about traffic in the locality, 25 per cent expressed dissatisfaction about the lack of facilities locally, and just under one fifth said that they disliked too much development.

Class position and spatial location shape the geographies of childhood in very precise ways. Despite the characterisation of life in Ratoath as peaceful, quiet and rural, there is a counter-discourse that expresses some anxiety in relation to the safety of children and the impact of traffic and poor infrastructure on quality of life. Parents residing in one local estate must drive their children the one hundred meters distance to school because there is no safe footpath that the children can traverse. Numerous accidents have occurred in the locality. Older pedestrians and young children are seen as particularly vulnerable. Older respondents report that they 'can't cross the road' anymore and that it is 'dangerous to leave your house' as a result. Children in the local national school wrote of their own fears in relation to their road safety and repeatedly made reference to traffic accidents in the locality.

> Ratoath has turned into a very busy place. My Mam and Dad told me, years ago you could play on the roads and only one car would pass every day. I've made a lot of friends but I think that Ratoath has got too busy, and sometimes I wonder why do all these people want to live in Ratoath? I do like having lots of friends but the roads are really busy now, and people can get killed really easily! (Sixth Class pupil, Ratoath NS).

The awareness of and sensitivity to risk shapes the discourses that emerge around children and childhood (Bonner: 1997, 121). Ex-urbanites in Bonner's study in the Canadian town of Prairie Edge see the benefits of moving out of the city in terms of the higher visibility of their children, the mutuality of guardianship in the community and the greater safety. The possibility of safety for the children of Ratoath is somewhat negated by the poor planning that has resulted in heavy traffic volume through the village, and the lack of provision for pedestrians and cyclists.

The greater dependency generally on cars impacts negatively on social lives. Several members of the Active Retirement group explained how they use to socialise in a local hostelry as well as outside the parish church after Sunday Mass. These social activities ,however, had ceased to be practised in the same way or to the same extent, because they felt they 'could not be delaying anyone'. The liberty they had previously felt to stop and interact with friends and neighbours had been stifled by their dependency on getting lifts from car owners. This had greatly diminished their 'chances to chat', an important element of social capital.

The autonomy of local children is also compromised by their dependence on being ferried about in cars. There is virtually no public transport serving Ratoath, and aside from a limited number of activities held in the village community centre, there are no amenities for young children and teenagers. To avail of almost any recreational activity they must be driven elsewhere by parents. Since there is no secondary school in the locality, all of the children will be attending schools in places that are not accessible other than by school bus or parents' cars. Paradoxically, children's horizons of opportunity are expanded greatly because of their car mobility, but their freedom of movement locally within the environs of the village are relatively restricted.

Surveillance is much more important (and occurs primarily through the medium of mobile phones) in an era when there is much greater awareness of risk. While moving out of the city and into a rural location has eased parents general anxieties about the safety and well being of their children, the latter cite safety reasons as the key factor in personal mobile phone ownership. This suggests that while objectively it is recognised that a country setting is 'safer' than a city neighbourhood, subjectively, the local population have ongoing worries about the safety and security of their own individual child. Hence, the proliferation of mobile phones.

On the positive side, the rapid expansion of Ratoath has specific benefits for both adults and children. Local children were adamant that Ratoath was a place where it was easy to fit in and make new friends. People 'don't judge you by your background' and the word openness was also used to denote the sense of Ratoath being a nice friendly place in which to live. Just under half of the people living in the locality have arrived only in the last five years. Forty per cent of the population sample express a preference for socialising in Ratoath, with the next largest proportion, 24 per cent, preferring to socialise at home. This indicates that people have developed a strong local orientation, and are much more likely to remain in and around the locality than to seek sociability outside of it. Some respondents suggested that as a developing suburb, newcomers were 'all in the same boat' and were therefore more aware of extending the hand of friendship than one would perhaps normally find in other communities.

As far as children are concerned, the development of the community has the great advantage of bringing more children into the locality thus widening their social circles. Children typically provide an important focus in Ratoath both because of their own interactions with other children living close by (a form of neighbourliness in itself) and because they draw their parents into contact with the parents of these other children. Networks of parents structured around the circulation spaces of children, in fact are among the most common and stronger forms of neighbourhood micro-communities. In Ratoath, these micro-communities exist in the immediate locales of the estates which are relatively autonomous from each other. The physical design of some estates makes for greater potential of sociability than others. Thus, estates that are designed around a common green, are more likely to create opportunities for sociability through, for example, football games or Halloween night activities. Those that are designed in the form of small cul de sacs provide less opportunity for creating common areas of communal activity.

Technology in the Community

According to Wellman (1999), the most significant social relationships are said to occur increasingly outside residential neighbourhoods. Local communities are being replaced by 'personal communities'. The networks of support and social ties that count are widely dispersed spatially: they focus on family, kin, friends and, some would add, on emerging 'virtual communities', sustained not through social interaction but through computer mediated exchanges. Wellman and Hampton (1999) see a shift occurring from living in little boxes (suburban homes) to living in networked societies, where boundaries are more permeable, interactions occur with diverse others, and linkages switch between multiple networks.

As a community, Ratoath is very much tuned into information and communications technology. More than half of households have three or more televisions. Although a relatively new technology, slightly more than half of households own a DVD machine. Only a handful of respondents do not possess a mobile phone, with the majority of households boasting two or more. Indeed, the vast majority of the sixth class children who attend the local primary school, reported that they personally own mobile phones. They primary reasons they cited for ownership was safety and being able to contact their parents in an emergency.

More than 80 per cent of households are equipped with a Personal Computer, and a further 30 per cent also own a lap top computer. Three-quarters of all households are connected to the internet. Both statistics are significantly above the national average: in 2000, about one third (32.4 per cent) of all households in Ireland had a home computer, and about 63 per cent of those households were connected to the internet (*Quarterly National Household Survey, 2000*).

In Ratoath, we found that personal computers are primarily used to access information, for e-mail and for word-processing. About 40 per cent of respondents also use their computer to play computer games. Just over a quarter of respondents (29.6 per cent) bank on-line, while almost the same proportion (27.7 per cent) shop on line. The relatively low level of take up of e-commerce services suggests that even in 'wired communities' people prefer to conduct their business in more traditional ways. Despite all the hype about new technologies and the interactive possibilities they offer, most respondents (53.7 per cent) cite the television as their preferred medium. The personal computer comes in second (preferred by 12. 9 per cent of respondents) followed closely by the mobile phone (12. 4 per cent). The most frequently cited reasons for keeping an item was it's entertainment value and usefulness as an information and communication tool. It is noteworthy that for those who use e-mail – about two-thirds of the respondents – the main purposes to which it is put is to stay in touch with family and friends. The potential of residents of Ratoath to develop strong linkages and identification with 'virtual' communities is high, given their ready access to information and communication technology. However, the evidence from our survey suggests that they use this technology in a limited number of ways. Rather than extending their linkages across far-flung kinship, workplace, interest groups and neighbourhood ties, they use e-mail to reinforce family and friendship bonds. Their use of computer networks therefore appears to enhance connectivity across pre-existing networks, rather than fomenting new networks in cyberspace. The data suggests that, in the case of Ratoath, on-line communities are not replacing in-person encounters, nor are they pulling people away from meaningful household and neighbourhood interactions.

Nevertheless, three-quarters of respondents see technology generally as allowing them to be more flexible about keeping in touch with people, and accessing places and communities other than Ratoath. While slightly more than two-thirds of people feel happy about the new communications technology in their lives, about a third say they sometimes feel overwhelmed and wish they could get back to a more simple way of living. Two-thirds of those surveyed also believe that there is excessive hype about new information technologies that make them seem more important than they really are.

Conclusion

In Ratoath, there is evidence of the erosion of the particularity of place as a result of its increasing size. Development has brought new life and livelihood (there are 38 businesses and services in the village) to the area, but the locale of Ratoath – a setting in which social relations are constituted – has been overwhelmed by the sheer demographic pressure brought to bear in recent years. The landscape is altering, as estates colonise the fields and woodland, removing the few remaining free play areas available to children. To avail of necessary services and amenities, Ratoathers must commute between other similar outposts on the city fringe. People do not 'go into town' anymore but rather are spatially mobile across the out of town commuter zone.

Children's friendship networks are estate-based, largely because of the difficulties (transport) and dangers (traffic) of moving away from their own turf. The new estates provide them with potential new friends. The estates make you 'city people' one youngster observed. On the other hand, those children who did not reside in housing estates but in more typically rural detached houses identified themselves more definitely with the county of Meath and not Dublin. In general, there was considerable confusion as to whether Ratoath was part of the countryside or part of the ever-expanding city. The existence of many houses, in particular, seemed to be the main criterion upon which defining a place as either rural or urban was based. An absence of houses meant countryside. Those who had moved from inner suburbs of Dublin rejected the epithet of 'boggers' assigned by friends from their former neighbourhoods. They identified strongly with their estate and therefore with a townie identity.

Bonner (1997) suggests that Prairie Edge, a Canadian suburban community, had the best of both worlds – the amenities and facilities of a city but the character and personality of a small town. Residents tended to describe the place positively (that it has both urban and rural characteristics) rather than negatively (that it is neither country nor city, neither urban nor rural). The term *rurban* has been coined to identify the settlement type that has both rural and urban characteristics , and thus ...'it is an appropriate way of indicating the particularity of the Prairie Edge difference. People in the locality associate rurality with notions of safety, of convenience, or reduced parental anxiety, and of high visibility' (Bonner: 1997, 111). So is a place like Ratoath *rurban*? Our data would seem to suggest not. Notions of rurality for respondents in the focus groups are tied to the aesthetic of the countryside, and the notion of a friendly, warm community. But Ratoath does not necessarily deliver the real and practical benefits of safety, convenience and reduced parental anxiety. On the other hand, people feel a strong identification with the place where they live. Spatial proximity fosters a sense of community which is real and vibrant, and which does not appear to be adversely affected by the widespread access to information and communication technology.

Bibliography

Boh, Katja, (1989) 'European Family Life Patterns – A Reappraisal' in *Changing Patterns of European Family Life* (edited by K. Boh *et al*), pp. 265-298, (London: Routledge).

Bonner, Kieran, (1997) *A Great Place To Raise Kids: Interpretation, Science and the urban-rural divide*, (Canada: McGill Queen's University Press).

Corcoran, M. P., (2000) 'Mall City' in M. Peillon and E. Slater (eds), *Memories of the Present*, (Dublin: IPA).

Corcoran, M. P., (2002) 'Place attachment and community sentiment: A European case study' in *Canadian Journal of Urban Research*, Vol 11, No 1, Summer.

Diani, Mario, (1995) *Green Networks: A Structural Analysis of the Italian Environmental Movement*, (Edinburgh: Edinburgh University Press).

Duncan, J. and D. Ley (eds), (1993) *Place, Culture, Representation*, (London: Routledge).

Edwards, B. and M. W. Foley, (2001) 'Much Ado about Social Capital' in *Contemporary Sociology*, Vol 30, No 3, May.

Fischer, C. S., (1975) 'Toward a Subcultural Theory of Urbanism' in *American Journal of Sociology*, 80: 1319-1341.

Fischer, C. S., (1982) *To Dwell Among Friends: Personal Networks in Town and City*, (Chicago: Chicago University Press).

Fishman, Robert, (1987) *Bourgeois Utopias: The Rise and Fall of Suburbia*, (New York: Basic Books).

Gans, H. J., (1982) *The Urban Villagers*, (New York: The Free Press).

Gans, H. J., (1962) 'Urbanism and Suburbanism as Ways of Life: A Re-evaluation of Definitions' in *Human Behaviour and Social Processes: An Interactionist Approach*, (edited by A. M. Rose), pp 625-648, (Boston: Houghton Mifflin).

Garreau, J., (1991) *Edge City: Life on the New Frontier*, (New York: Doublday).

Hummon, David, (1992) 'Community Attachment: Local Sentiment and Sense of Place' in *Place Attachment* (edited by I. Altman and S. M. Low) pp 253-276, (New York: Plenum Press).

Humphreys, Alexander, (1966) *New Dubliners: Urbanisation and the Irish Family*, (London: Routledge and Kegan Paul).

Jacobs, Jane, (1961) *The Death and Life of Great American Cities*, (New York: Vintage).

Ladd, Everett, (1999) *The Ladd Report*, (New York: Free Press).

Melucci, Alberto, (1996) *Challenging Codes: Collective Action in the Information Age*, (Cambridge: Cambridge University Press).

Quarterly Household Survey 2000, (Dublin: Central Statistics Office).

Relph, J., (1976) *Place and Placelessness,* (London: Pion).

Sennett, Richard, (1970) *The Uses of Disorder: Personal Identity and City Life,* (New York: Random House).

Simmel, G., (1960) *The Metropolis and Mental Life in Image of Man: The Classic Tradition in Sociological Thinking,* pp 437-448 (edited by C. W. Mills) (New York: George Brazillier).

Stack, Carol, (1974) *All Our Kin: Strategies for Survival in a Black Community,* (New York: Harper and Row).

Tovey, Hilary, (2002) 'Rethinking Urbanisation: struggles around rural autonomy and fragmentation' in M. P. Corcoran and M. Peillon (eds), *Ireland Unbound: a turn of the century chronicle,* (Dublin: Institute of Public Administration) pp 167-185.

Varenne, H., (1993) 'Dublin 16: Accounts of Suburban Lives' in *Irish Urban Cultures* (edited by C. Curtin, H. Donnan and T. M. Wilson) pp. 99-122, (Belfast: Institute of Irish Studies).

Wellman, Barry (ed), (1999) *Networks in the global village,* (Boulder Col.: Westview Press).

Wellman, B. and K. Hampton, (1999) 'Living Networked On and Off Line' in *Contemporary Sociology,* Vol 28, No 6, pp 648-54.

Wirth, Louis, (1938) 'Urbanism as a Way of Life' in *American Journal of Sociology,* 40: 1-24.

Young, Michael and Peter Wilmott, (1957) *Family and Kinship in East London,* (London: Routledge and Kegan Paul).

CHAPTER TWELVE

The Relevance of Early Heidegger's Radical Conception of Transcendence to Choice, Freedom and Technology

Paul Downes

The technological dystopia of Huxley's *Brave New World* (1932) envisaged the replacement of lovers with laboratories in the creation of children; it postulated a happiness pill 'soma' which society took as the epitome of its freedom, with its central female character Lenina cheerfully observing that 'We're free, everybody's free nowadays', while never experiencing the pain of insight as to the limits of her freedom. The only avenue for resistance to the hegemony of dull emotionless technocracy in *Brave New World* was the character of the savage, the last vestige of natural human being. Nature, in the form of the savage, was Huxley's site for raging against the dying of the light of human existence, a light submerged by the distilled shadow of technology.

Simple romantic subjectivity places expression of an inner natural world in tension with an inauthentic repressive cultural world, the poles of a tension within which Huxley's savage is caught, where technology is the vehicle and background force expressing cultural repression. Glorification of the natural is a position commonly attributed to Rousseau (e.g. Leach: 1976) though somewhat simplistic both of Rousseau (1762, 195-196), and of romanticism itself. It is arguably a framework for liberation relied upon by the humanistic psychology of Carl Rogers (1974) emphasising expression of an inner nature, and the existential work of R. D. Laing which criticised societal normality for 'con[ning] children out of play' (1967, 55) and betraying our true potentialities. Even though Freud was committed to deterministic explanations and resisted a romantic view of nature, he still sought to mediate between an expressive instinctive 'id' and a repressive cultural 'superego' – the Freudian ego was still caught within a nature/culture dualism.

Early Heidegger (1927, 94) postulated a more fundamental realm than that of the natural, namely, primordial truth and experience, while also resisting the classical position of Aristotle, Cicero, Aquinas, Locke, and Grotius *inter alia* (see, Kelly: 1992) that inner nature gains its expression in society. This primordial domain, expressed through early Heidegger's radical conception of transcendence, contrasts with later Heidegger's etymological approach to understanding of being (Levinas: 1986 / 1990, 152), where he more explicitly discusses the relation between technology and being. Early Heidegger boldly emphasised that his conception of transcendence in *Being and Time* is radically different from traditional approaches:

> [a] more originary conception of...transcendence, a basic determination of *Dasein's* whole existence...[is] a central problem that has remained *unknown to all previous philosophy* (Heidegger: 1927 / 1982, 162) (my italics)

The difference with later Heidegger is stark:

> The preparatory thinking in question ... only attempts to say something to the present which was already said a long time ago precisely at the beginning of philosophy... (Heidegger: 1973, 60)

Early Heidegger (1927) emphasises transcendence, whereas later Heidegger focuses more on language and poetry in their relation to being.

The following interrelated questions will be examined in this essay predominantly with regard to *early* Heidegger rather than the often dubious (Tugendhat: 1991, 81; Jameson: 1990,16) interpretation of ancient Greek texts prioritised by later Heidegger: Why is technological thinking limited? How can early Heidegger's conception of primordial truth as transcendence offer a different basis for freedom than products of technocratic thinking?

Inauthentic Grounds to Fundamental Choice in Computer Models and Other Cognition-based *Forms of Choice*

Newell & Simon's (1972) General Problem Solver 'became practically the Bible of information processors' (Hearnshaw: 1987, 275). Later developed into Newell's (1990) Soar problem-search to accommodate domain-specific knowledge, such computer models of the mind adopt a metaphor of *search* in a problem space. Newell, a leading cognitive scientist in a currently dominant paradigm in psychology, describes Soar as a candidate cognitive architecture – a supposedly general framework for characterising all of cognition. For Newell, 'search is fundamental for intelligent behaviour. It is not just another method or cognitive mechanism, but a fundamental process' (1990, 96). Later Heidegger explicitly challenges search in 'the rigid measures' (Heidegger: 1959, 183) of such an abstract space:

> All becoming is regarded as motion and the decisive aspect of motion is change of place. With the rising domination of thought in the sense of modern mathematical rationalism no other form of becoming besides motion as change of place is recognised. Where other phenomena of motion show themselves, one tries to understand them in terms of change of place (Heidegger: 1959, 195).

Yet the need for a more primordial space is already evident in early Heidegger:

> ... the bare space itself is still veiled over. Space has been split up into places ... the primordial spatiality of Being-in is concealed (Heidegger: 1927, 139-141).

Heidegger would predict that the most important choices made by computer models, based on search as movement through an abstract space, are inauthentic (despite the emphasis in cognitive science that these models 'work' in fulfilment of pragmatic goals). Newell recognises that dynamic choice at lower sub-goal levels is far less important than dynamism at higher goal levels:

> Any change in a goal at some level – a change in problem space, state or operator – completely preempts all changes about subgoals, because these latter were all created in response to the impasse at this higher level (Newell: 1990, 175).

He describes a 'conflict impasse' at these highest levels:

> An impasse does not mean that Soar is stymied and therefore halts. Soar responds to an impasse by creating a subgoal to resolve it ... it must include some preferences that make the decision occasion that produced the impasse now yield a *clear* choice. *This does not require any formulation of the set of possibilities for what could resolve the impasse. The decision procedure itself determines that* (Newell: 1990, 176-77) (my italics)

The highest level choices are not resolved by any logical principle except a pragmatic one of clarity – a clear choice to enable action. A clear decision is by no means an intellectually principled decision as Hart (1961) recognises in the context of legal logic when characterising clarity as a secondary and not primary norm. While clarity to resolve choice could be justified to resolve trivial lower level sub-goal choices, it is a clear limitation at higher levels of reasoning which condition all the subsequent lower level choices for the Soar model. In legal logic this attempt to ground choice in the overriding deciding factor of clarity for decision-making would be characterised as a letter of the law approach contravening the spirit of the law, namely, the underlying goals the law was designed to achieve.

Search for strictly logical grounds for choice is also questionable according to implications of Gödel's theorem transferred from the context of mathematics to all logical systems. Hearnshaw (1987) suggests that Heidegger's preoccupation with a realm of Being prior to thinking may have anticipated Gödel's concerns. Hofstadter paraphrases Gödel's theorem: 'All consistent axiomatic formulations of number theory include undecidable propositions' (1979, 17). The more consistent a formal system, the less complete it is; the more complete it is, the less consistent it is. Though Hofstadter emphasises the differences between formal mathematical systems and logical systems reliant on language (1979, 51-2), and Sokal & Bricmont highlight abuse of Gödel's theorem in the social sciences, emphasising that at most it can be suggestive rather than demonstrative (1999, 169), the basic idea that no logical system contains its own proof has been independently accepted by leading twentieth century jurisprudential thinkers, Hart and Kelsen, with regard to legal logic. Hart's 'rule of recognition' determining the criteria which govern the validity of the rules of the legal system is a recognition of the inevitability of value judgements at the root of legal logic; as an ultimate rule it can 'neither be valid nor invalid' (Hart: 1961, 105). Kelsen's *Grundnorm* or basic norm grounding legal authority is similarly recognised as being prior to logical derivation:

> The document which embodies the first constitution is a real constitution, a binding norm, only on the condition that the basic norm is presupposed to be valid ... The basic norm ... is not – as a positive legal norm is – valid because it is created in a certain way by a legal act, but is valid because it is presupposed to be valid (Kelsen: 1946, 116-7).

The idea that truth or logic are not at the root of truth or choice statements is echoed both by Kuhn's (1970) description of a scientific paradigm shift as being influenced (whether primarily or secondarily) by extra-scientific, social considerations, and the postmodern suspicion of truth claims as being based not primarily on reason but power (Foucault: 1972).

The displacement of presence within logical grounds for choice implicates cultural power as the guiding force motivating choice, and products of technological thinking are a danger to society in that they provide no guide in themselves as to their own use as they do not ground themselves, while being very susceptible to the distorting influences of the desires of those with power in a culture. Heidegger locates the distorting influence of power in a culture not necessarily in a specific person or institution but in an impersonal 'they' self distinguished from the 'authentic self' (1927, 167):

> The 'they' has always kept *Dasein* from taking hold of these possibilities of Being. The 'they' even hides the manner in which it has tacitly relieved *Dasein* of the burden of explicitly choosing these possibilities. It remains indefinite who has 'really' done the choosing. So *Dasein* makes no choices, gets carried along by the nobody and thus ensnares itself in inauthenticity. This process can only be reversed if *Dasein* brings itself back to itself from its lostness in the 'they' (Heidegger: 1927, 312)

The danger is that technology not so much constitutes the they self but is an increased impersonal expression of its force and values through (a) making the impersonal they self more concrete yet (b) keeping its value ladenness more hidden. Agency is taken away through the 'dictatorship' (1927, 164) of the they self, as choice is in effect surrendered to impersonal cultural forces:

> In *Dasein's* everydayness the agency through which most things come about is one of which we must say that 'it was no one'... (Heidegger: 1927, 165).

Heidegger's criticism of choice surreptitiously controlled by the they self or cultural forces, anticipated criticisms made of subjectivist or constructivist agency. According to Williams (1992), Heidegger offers an alternative to the standard view of constructivist agency in psychology which treats choice as that of choosing between alternatives. Constructivist choice is open to the criticism that it fails to clarify the source of the criteria or grounds upon which a choice is made. The constructivist position fails to resolve the problem of an infinite regress of choices concerning criteria for choices, and is open to the danger that the criteria upon which a choice is made is not created by the individual but by cultural forces (Garfinkel: 1967; Williams: 1992). Cognitive or constructivist choice threatens to be the choice of a 'cultural dope' (Garfinkel: 1967) who fails to see the abdication of freedom in the very act of choice akin to Lenina's vacuous freedom in *Brave New World*. Heidegger's (1927) criticism of the Cartesian *cogito ergo sum*, namely that truth and experience grounded in thought is unprimordial, is echoed by such criticisms of constructivist agency. From this perspective, products of technological thinking are simply as vulnerable to the criticism of being unprimordial as any other grounds for experience founded in cognition. The technocrat is little different from the cognitive therapist who tries to convince a client to restructure his or her thinking from 'negative' to 'positive' thoughts, as the new 'positive' thoughts to help 'adjustment' are simply an alternative mouthpiece from the same culture to adjust to that culture (see also Fernando: 1991; Marsella & White: 1982 on Western biased assumptions in therapy and conceptions of self).

Pragmatist readings of Heidegger (Guignon: 1983, Dreyfus: 1991, Rorty: 1991) fail to accommodate early Heidegger's search for a fundamental ontology, as they reject even the possibility of a primordial truth realm.[1] A pragmatic credo that what is useful is what counts remains blind to the goals underlying such utility and the process of selection and prioritisation of certain goals over others. Utility presupposes clarity of the goal(s) selected and consensual criteria for assessment of results but does not explain the origins of selection of this goal over other possible goals. Similarly, Harré's discursive psychology delimits its domain of relevance to avoid fundamental issues of an agency resisting cultural manipulation:

> what I am trying to sketch is not a psychology of choosing, but one of executing a choice once made (Harré: 1995, 120)

Pragmatists neither succeed nor try to offer view of the goals of utility which transcend the reference point of common sense within a given culture. Eagleton's critique of postmodern pragmatism equally applies to constructivist choice and computer modelled choice such as Newell's Soar:

> Interests and desires operate, in effect, if not in admission as quasitranscendental anteriorities; there can be no asking from where they derive ... [they] are simply withdrawn from the process of rational justification, as that which one can never get behind (Eagleton: 1990, 379-383).

Adapting T.S Eliot's words, this unprimordial choice is the creation of a dry mind, its criteria for choice being the dry season of postmodern and computer modelled pragmatism. Early Heidegger's quest is for a transcendence that goes beyond transcendence *qua* abstraction, beyond 'empty logical possibility' (1927, p.183) and transcendence as a 'ground' for truth.[2]

As yet, the main focus has been on what transcendence as freedom is *not* for early Heidegger. While culture *qua* abstraction from nature has been characterised as a potentially oppressive force, namely, a 'dictatorship' of the they self with technology as one of its instruments, potential transcendence is rooted for Heidegger neither in nature nor mind as abstraction. Heidegger is not dining from the same menu as either Huxley or Descartes. Moreover, computer models' movement through an abstract conceptual space is an inauthentic mode of being for Heidegger so that transcendence requires a different spatial quality. However, he is not content with truth or experience as a ground, so this different mode of spatial relation would not be a ground as such. Early Heidegger's transcendence was a horizon rather than ground of transcendence, in other words, a conception of transcendental truth as a direction more than position – a view supported by 'the subversion of the traditional metaphysical priority of actuality over possibility' (Kearney: 1994, 299) in *Being and Time*. The horizon of transcendence does not have the certainty of a Hegelian *telos* (Gillespie: 1984, 153; Guignon: 1983, 80).

Heidegger's freedom as transcendence is neither a constructivist nor pragmatic agency reducible to becoming merely a 'cultural dope' of the they self, where technological products of instrumental rationality are radically divorced from sensitivity to desire and values. While these criticisms of technology echo criticisms of a 'letter of the law' approach to legal logic and parallel Weberian criticisms of the tyranny of impersonal bureaucratic rationalism, is there a further level to Heidegger's criticisms of technology implicit in his early radical conception of a horizon of transcendence?

Background Structures of Relation Framing Choice: Assumed Connection and Separation

Though rejecting constructivist agency, Williams (1992) treats morality as the ground for choice for Heidegger. Yet this role of morality is clearly inconsistent with Heidegger's rejection of 'grounds' for choice (see also Dreyfus: 1993, and Norman: 1997 on the foundational metaphor) and Williams also fails to offer any account of how this morality is to transcend the cultural conditioning within which the constructivist agency he criticises is trapped. It is important to emphasise that the primordial truth domain expressing Heidegger's transcendence is a *unity* within experience and truth which is not identical with *all* experience or truth *totality* (Richardson: 1963, 74; Brooke: 1991, 96). Heidegger's concern is with a pre-cognitive realm of Being, his search is for 'conditions' and 'structures' of Being *framing* choice. This can be clarified through development of Gilligan's (1982) qualitative research contrasting two different relational approaches framing the moral problem solving of adolescent boys and girls.

According to Gilligan, one assumption or relational state framing moral reasoning is that of an *assumed connection* between self and other – an assumed connection of 'ethic of care' which Gilligan controversially suggests is 'associated' (1982, 2) with female moral reasoning. She contrasts this with an abstract, hierarchical, impersonal 'logic of justice' approach based on a prior relational state of assumed separation between self and other (the approach implicitly prioritised by Kohlberg's (1958, 1981) examination of moral reasoning). For example, Gilligan contrasts two children's modes of thought and relation:

> To Jake, responsibility means *not doing* what he wants because he is thinking of others; to Amy, it means *doing* what others are counting on her to do regardless of what she herself wants ... she, assuming connection, begins to explore the parameters of separation, while he, assuming separation, begins to explore the parameters of connection (Gilligan: 1982, 38).

These relational states of assumed separation and connection frame moral thinking and choices (see also Schon: 1993, Reddy: 1993, on problem framing), leading to differing emphases on an impersonal abstract balancing of rights and a contextual focus on relationships, respectively. For Gilligan, experiential relational states are prior to thought itself, echoing a phenomenological approach of experience being prior to the idea.

Recent interviews[3] with teenagers from Tallinn, Estonia and Vilnius, Lithuania, adopting a qualitative approach to examine various aspects of motivation, support a proposed view of agency as movement from culture bound constructivist choice relying on abstraction and information – also expressed in technocratic thinking – to a choice relying more on a connective framing mode or structure of relation. Eve, a 14 year old Estonian girl gave the following response to the question 'Do you understand yourself ?':

> You mean, do I trust myself ? Of course, I do. No, you actually cannot trust yourself like that. Sometimes you don't know yourself whether you can trust yourself or not – sometimes you have that kind of feeling.

Understanding or knowledge of the self is described in terms of a relationship of trust – of assumed connection to oneself. Eve reformulates or reframes the question to fit in with her experiential schema of assumed connection. Jurga, a 14 year old Lithuanian girl gave the following response to the question 'Do you mostly rely on yourself or on others ? Which do you think is the best approach ?':

> At first you have to trust yourself, and when you trust yourself, you trust others as well. I trust myself and others.

Jurga reconstructs a self/other dichotomy into an assumed connection between self and other which is not simply equating self and other. The initial question is framed within either/or, Aristotelian logic categories of A (self) and non-A (other). Jurga reconstructs the question frame from traditional Western logic through her assumed connection logic of experience. Merle, another 14 year old Estonian girl offered the following response to the question 'Do you mostly trust yourself or others ?':

> With like making choices ? [When told yes, she continued] Then I maybe trust more ... in the beginning I trust the others, listen to their opinions and then I like start to push on my will. So, I like trust more myself.

Merle expressly links trust, whether of self or other, with choice. She interprets the question concerning trust, i.e. connection or relation, as being consistent with her experiential schema of making choices. She intuitively relates choice to relation, trust of self or others. There was nothing in the initial question concerning trust which required her to formulate her answer in terms of the context of 'making choices'.

Understanding, choice and relation of self to others are governed by prior experiential relational states of assumed connection or trust for these teenage girls. Choice takes place not against a neutral background but fundamentally in relation to a process of trust or lack of trust in self and/or other. The assumed connection schemas reframe moral problems, not through a generic formula, but through a relation of trust which includes the anxiety of uncertainty. For Heidegger (1927, 232) anxiety is central to authenticity – an anxiety and trust beyond the scope of technology. This movement from criteria for choice based on abstraction and information is also a movement from impersonal quasiautonomous abstract values towards an emphasis on an experiential capacity for valuing, which presupposes a basic identity of trust or assumed connection to what/who is to be valued. Loss of relation to self implicates a loss of relation to that which the displaced self then values; existential distance from self is a distance from all else the self connects with, including values. Gilligan (1990) refers to experience of this loss of relation through being swamped in Western cultural logic – encompassing technological rationalism – and antithetical to an understanding of self and relation:

> The either/or logic that Gail was learning as an adolescent, the straightline categories of Western thinking (self/other, mind/body, thoughts/feelings, past/present) and the if/then construction of linear reasoning threatened to undermine Gail's knowledge of human relationships by washing out the logic of feelings (Gilligan: 1990, 18-19).

Transcendence of Cultural Conditioning: Prior Dimensions of Space and Time Underlying Relations Framing Choice

Gilligan and Heidegger share a focus on experiential background conditions framing (moral) choice. Yet whereas Gilligan's (1982) assumed connection emphasises a relational state of empathy, Heidegger suggests that empathy is not primordial but derives from a prior connective level of 'Being-with' (1927, 162) – a prior level not simply reducible to humanistic subjectivity. The prior level of Being is now understandable as the background relation itself between assumed connection and assumed separation. Focus on such a background relation preceded *Being and Time* and was already a key question of Heidegger's doctoral thesis:

> What we need is a structure that is not merely a thinking together of the two surface structures of *synthesis* and *separation*, but which, being *a unitary structure, precedes both* (Heidegger: 1913/1978, 21) (my italics).

This transcendental level is for early Heidegger an *a priori* structure, 'the structure of existentiality lies *a priori*' (1927, 69). Understood as prior background structures of relation, namely, conditions of assumed connection and separation which frame choice, and their possible dynamic interrelation, *modes of being of assumed connection and separation assume a primordial spatial dimension within which connection (synthesis) and separation occur.* Assumed connection as trust includes a spatial mode of open relation, while terms such as expression and repression also assume prior structural modes of openness and closure respectively. From this perspective, the potentially dynamic *interaction* between these spatial modes of existential relation presupposes a dimension of temporality – a primordial temporality which is not simply reducible to linear historical time but is consistent with the temporality of Heidegger's (1927) transcendental horizon as being prior to existential spatiality (though see Malpas: 2000).

Relation presupposes a spatial realm, change in relation presupposes a temporal dimension. Different modes of relation affect background frames for choice. Opening of these different relations requires for early Heidegger an opening of space and time that differs from linear historical time upon which socioculturally constructed reality is based. This opening is not a ground, but an uncertain possible horizon. Whether airplanes, television, radio or e-mail, technology reconstructs our relationship to physical and communicative space. Technological increases in the efficiency of communication, the accentuated capacity for rapid transfer of information, as well as the reorientation of experience of painful memories through chemicals such as Prozac (perhaps the modern day version of Huxley's 'soma'), are significant alterations of our experience of time. Early Heidegger[4] would characterise technological restructuring of space and time as inauthentic, to the extent that it relied on the mode of being of assumed separation of which abstraction from experience is one form. Changing the angle of relation in space through a transcendent temporal dimension is an increased direction towards assumed connection – a capacity alien to the desiccated frames of technology. Primordial experience is the unifying process between experientially split modes of assumed connection and separation.

Endnotes

1. However, one of the most prominent advocates of a pragmatist reading of *Being and Time*, Dreyfus (1991), has recently stated that he accepts that Heidegger was not a pragmatist (Dreyfus: 2000, 333). The reductionist approach of Dreyfus (1991) to Heidegger's text has been noted *inter alia* by Dahlstrom (1995), Norris (1997) and Haapala (1997).
2. Later Heidegger similarly criticises pragmatic justifications of technological rationalism as lacking clarification of underlying goals: 'The technological scientific rationalisation ruling the present age justifies itself every day more surprisingly by its immense results. But these results say nothing about what the possibility of the rational and the irrational first grants' (1973, 72).
3. Supervised by the author with the assistance of funding from the Research Committee, St Patrick's College, Drumcondra, and a research assistant from Concordia International University, Estonia.
4. Later Heidegger explicitly characterises the accelerated temporal calculations of computers as being derivative: 'Today, the computer calculates thousands of relationships in one second. Despite their technical uses, they are inessential' (Heidegger: 1957, 41).

Bibliography

Brooke, R., (1991), *Jung and Phenomenology*, (London: Routledge).

Dahlstrom, D. O., (1995), 'Heidegger's Concept of Temporality: Reflections on a Recent Criticism', *Review of Metaphysics*, 49, 95-115.

Dreyfus, H. L., (1991), *Being-in-the World: A Commentary on Heidegger's Being and Time, Division I*, (Cambridge, MA: MIT Press).

Dreyfus, H. L., (2000), Reply to William Blattner, in, *Heidegger, Authenticity and Modernity: Essays in Honour of Hubert L. Dreyfus. Vol.1.*, Wrathall, M., & Malpas, J., (eds.). (London & Cambridge: MIT Press).

Eagleton, T., (1990), *The Ideology of the Aesthetic*, (Oxford: Blackwell Publishers).

Fernando, S., (1991), *Mental Health, Race & Culture*, (Basingstoke: MacMillan).

Foucault, M., (1972), *The Archaeology of Knowledge and the Discourse on Language*, trans. Smith, A.M. Sheridan, (New York: Pantheon Books).

Garfinkel, H., (1967), *Studies in Ethnomethodology*, (Englewood Cliffs NJ: Prentice Hall).

Gillespie, M. A., (1984), *Hegel, Heidegger and the Ground of History*, (Chicago: University of Chicago Press).

Gilligan, C., (1982), *In a Different Voice*, (Harvard, CT: Harvard University Press).

Gilligan, C., Lyons, N. P., & Hanmer, T. J., (eds.), (1990), *Making Connections*, (Harvard, CT: Harvard University Press).

Guignon, C. B., (1983), *Heidegger and the Problem of Knowledge*, (Indianapolis: Hackett Publishing Co.).

Haapala, A., (1997), 'Interpreting Heidegger Across Philosophical Traditions', *Metaphilosophy*, 28, 433-48

Harré, R., (1995), 'Agentive Discourse', In Harré, R., & Stearns, P., (eds.), *Discursive Psychology in Practice*, (London: Sage Publications).

Hart, H. L. A., (1961), *The Concept of Law*, (Oxford: Oxford University Press).

Hearnshaw, L. S., (1987), *The Shaping of Modern Psychology*, (London-New York: Routledge & Kegan Paul).

Heidegger, M., (1913/1978), *Die Lehre vom Urteil im Psychologismus: Ein Kritisch-Positiver Beitrag zur Logik*, Dissertation, Freiburg in Br. Reprinted in, Heidegger, M., *Gesamtausgabe, Bd 1*, (Frühe Schriften Frankfurt am Main: Klostermann).

Heidegger, M., (1927), *Being and Time*, trans. by MacQuarrie, J., & Robinson, E., (1962). (Oxford: Basil Blackwell.).

Heidegger, M., (1927/1982), *Basic Problems of Phenomenology*, (Bloomington: Indiana University Press).

Heidegger, M., (1957), *Identity and Difference*, trans. by Stambaugh, J., (New York: Harper & Row).

Heidegger, M., (1959), *Introduction to Metaphysics*, (New Haven, CT: Yale University Press).

Heidegger, M., (1973), *The End of Philosophy*, (New York: Harper & Row).

Hofstadter, D. R., (1979), *Godel, Escher, Bach: An Eternal Golden Braid*, (New York:Basic Books).

Jameson, F., (1990), *Postmodernism or the Cultural Logic of Late Capitalism*, (London:Verso).

Kearney, R., (1994), *Modern Movements in European Philosophy: Phenomenology, Critical Theory, Structuralism*, (Manchester, UK: Manchester University Press).

Kelly, J. M., (1992), *A Short History of Western Legal Theory*, (Oxford: Clarendon Press).

Kelsen, H., (1946), *General Theory of Law and State*, (Cambridge, MA: Harvard University Press).

Kohlberg, L., (1958), *The Development of Modes of Thinking and Choices in Years 10 to 16*, Dissertation, (Chicago: University of Chicago)

Kohlberg, L., (1981), *The Philosophy of Moral Development*, (San Francisco: Harper & Row).

Kuhn, T., (1970), *The Structures of Scientific Revolutions*, 2nd edn. (Chicago: University of Chicago Press).

Laing, R. D., (1967), *The Politics of Experience*, (Harmondsworth: Penguin).

Leach, E., (1976), *Claude Levi-Strauss*, (Harmondsworth: Penguin).

Levinas, E., (1986), 'Admiration and Disappointment', in Neske, G., & Kettering, E., (eds.), *Martin Heidegger and National Socialism*, 1990, trans. by Harries, L., & Neugroschel, J., (New York: Paragon House).

Malpas, J., (2000), 'Uncovering the Space of Disclosedness: Heidegger, Technology and the Problem of Spatiality in *Being and Time*', in *Heidegger, Authenticity and Modernity: Essays in Honour of Hubert L. Dreyfus. Vol.1.*, Wrathall, M., & Malpas, J., (Eds.), (London & Cambridge: MIT Press).

Marsella, A. J., & White, G. M., (1982), *Cultural Conceptions of Mental Health and Therapy*, (Dordrecht-Boston: D.Reidel).

Newell, A., & Simon, H. A., (1972), *Human Problem Solving*, (Englewood Cliffs, N.J: Lawrence Erlbaum Assoc.).

Newell, A., (1990), *Unified Theories of Cognition*, (Cambridge, MA: Harvard University Press).

Norman, A., (1997), 'Regress and the Doctrine of Epistemic Original Sin', *The Philosophical Quarterly*, 47, 477-490.

Norris, C., (1997), 'Ontological Relativity and Meaning-Variance: A Critical-Constructive Review', *Inquiry*, 40, 139-73

Reddy, M. J., (1993), 'The Conduit Metaphor: A Case of Frame Conflict in our Language about Language', in Ortony, A., (ed.), *Metaphor and Thought*, (Cambridge: Cambridge University Press).

Richardson, W. J., (1963), *Through Phenomenology to Thought*, 3rd ed 1974, (Martinus Nijhoff: The Hague).

Rogers, C., (1974), *On Becoming a Person: A Therapist's View of Psychotherapy*, (London: Constable).

Rorty, R. (1991), *Essays on Heidegger and Others: Philosophical Papers Vol.II*, (Cambridge: Cambridge University Press).

Rousseau, J-J., (1762), *The Social Contract*, trans. by Cranston, M., 1968 (Harmondsworth: Penguin Books).

Schon, D., (1993), 'Generative Metaphor: A Perspective on Problem Setting in Social Policy', in Ortony, A., (ed.), *Metaphor and Thought*, (Cambridge: Cambridge University Press).

Sokal, A., & Bricmont, J., (1999), *Intellectual Impostures: Postmodern Philosophers' Abuse of Science*, (London: Profile Books).

Tugendhat, E., (1992), 'Heidegger's Idea of Truth', in Macann, C., (ed.), *Martin Heidegger: Critical Assessments*, (London-New York: Routledge).

Williams, R. N., (1992). 'The Human Context of Agency', *American Psychologist*, 47, 752-760.

CHAPTER THIRTEEN

For Heaven's Sake – What On Earth Does Technology Have To Do With Transcendence?:

A Theological Meditation on Kieslowski's *Dekalog 1*

Stijn Van den Bossche

> He casts forth his ice like morsels: who can stand before his cold? He sends forth his word, and melts them; he makes his wind blow, and the waters flow (Ps 147, 17-18).

Dekalog I, directed by Krzysztof Kieslowski and co-written with his friend Krzysztof Piesiewicz in 1988, is the first in a series of ten films based on the Ten Commandments. The main setting of all ten films is the Stawski quarter in Warsaw. My focus here will be on what *Dekalog I* reveals with regard to the relationship between technology and transcendence, or the lack thereof. Let me add one specific introductory remark: it seems to me that Kieslowski's depiction of the commandments marks him as not only a great artist, but as someone who understands quite well what Christianity is about even though he considers himself to be agnostic. He even seems to think very highly of Roman-Catholicism. This is not a kind of 'papolatry' which is often associated with traditional Polish Catholicism: Kieslowski recognises and apparently respects the metaphysical truth claims of Christian and even Catholic faith. These films therefore present a congruity between art and the biblical message. Indeed, the films might be considered a contemporary pastoral tool elucidating the Ten Commandments since they act as a hermeneutic of the Commandments. Allowing themselves to be 'commanded' by the message of the biblical Decalogue, the films are a truly artistic contemplation of the metaphysical truth of the Commandments rather than 'instruments' to explain the Decalogue. Kieslowski's *Dekalog* therefore, does not put revelation at the service of art, nor art at the service of revelation. For him, there is simply no distance between the two. It might rather be suggested that Kieslowski has made ten cinematic icons – each film offering a vision of each commandment which interacts uniquely with each individual viewer.

Our Basic Condition: Our Being of God

Dekalog I opens with a scene of ice merging into water, the perspective gradually broadening to reveal what appears to be a blow-hole in the ice. The camera then pans over the ice to reveal a man mysteriously contemplating a wood fire burning at the edge of a lake located near a snow-covered square in Warsaw. Nobody seems to take notice of him, but he is there and will remain there. As such, he is 'outside the narrative' of the film, and indeed of history – he is just being (there). Initially, he appears to be a homeless person, but there is also from a theological perspective a clear evocation here of the Burning Bush (Ex 3), the place where Moses meets God personally, the 'place' where history meets the otherness of history – the Transcendent. The first commandment acts as a 'sacred' canopy for the other nine: 'I am your God (...) You shall have no other gods apart from me' (Ex 20, 2-3) (Beauchamp: 1999, passim). Contemplating the fire as Moses did the burning bush, the man is 'the Human' whose existence depends on God, and perhaps in a particular way on his encounter and communion with God as the first condition of his being. The figure of the man acts as a reminder that everything that happens in (the) history (of the film) arises from within the basic framework of our being, being which is dependent on the Transcendent.

It is crucially important to recognise that, especially in his visual elaboration of the (Ten) Commandment(s), Kieslowski avoids being moralistic or accusatory. Instead, through the medium of his art, he portrays that which undeniably appears to be the truth about life. The technological tragedy which the film chronicles is not caused by the obliteration of transcendence. Instead, we are reminded that every technology that permits a form of mastery within life does not itself constitute a mastery of life. Being is utterly dependent on Transcendence and at the end of the day, technology, in itself, really has nothing to say about transcendence. It merely functions – beneficially or detrimentally – within a reality which is, as such, entirely related to the transcendent. This is not a (moral) judgement regarding technology; still less a condemnation. It is instead an observation about the ultimate truth of life despite the fact that such truth tends to be forgotten by both the characters in the film and the world in our own time. Technology is a good in itself, but to rely upon it to provide metaphysical foundation for one's life is a tragic mistake.

There are two close-ups of the man at the burning fire in the film's opening scenes. These provide a frame for an intervening image in which the camera zooms in on the character of Aunt Irena. Her name is not coincidental, the Greek *eirènè*, translating the Hebrew *shalom* – the Paradisiacal state which history reaches when it finds communion with God. *Shalom* is the peace which God's people shall reach when they are 'with God', and which, as the prophets repeatedly warn, will be lost if they neglect God. *Shalom* is the peace about which the angels sing in utmost joy when Christ is born. It is the peace with which Christ greets his disciples after his resurrection. It is the peace of the end-times for which we pray in the Eucharist and which we dare already to share with one another in Christ. Indeed, it is precisely because the Eucharist offers communion with God in Christ, that the Fathers of the Church called it the sacrament of unity and peace, or even more concisely, *Pax*. History shall find such peace when it adopts the approach of the man at the fire, when it grasps its relationship to God. Aunt Irena (Aunt Peace) is therefore the character who has internalised this attitude, the one who lets herself be framed by the reality of humanity's being-in-relation to God. She 'does' what the person at the fire 'is'.

Aunt Irena is observed watching a television set displayed in a shop window, on which there is a programme about children playing at school. One of the children is her nephew Pawel, filmed by the local television station for a health project. At this point the viewer is unaware of the connection, but this brief moment captures the essence of the story upon which Kieslowski builds the remainder of the film. As Aunt Irena gazes at the television, a tear fills her eye, slowly wetting her cheek. The film cuts back to the man by the fire who looks straight into the camera before returning his gaze to the fire. He gently rubs his eye, as if to wipe Aunt Irena's tear from his own face. History is hopeless: it cannot save itself. The only true hope for human history is God; otherwise it is lost.

The Household at Stake:
History Founding Itself – Technology in Control

The narrative proper begins with a pigeon – symbolising the messenger of peace, the Spirit – flying to the top of a high, grey apartment building to reach some bread crumbs which Pawel has put on the window-sill. The bird is seen straining to reach up into the tower of human selfhood to receive the crumbs from a father and child whose world is centred on their technological selves (the mother does not appear; we are told she 'may call before Christmas'). In this image, Kieslowski would seem to infer the great humility of God. Though we do not look for God, God comes to look after us, and is content to let us think that he depends on us, accepting our crumbs as a gift. And God's strategy works, for we see that the child is genuinely happy when the pigeon comes.

We come across this incomplete human family as they are engaged in quite difficult physical exercises. In their self-understanding the body is the self, and therefore taking 'technological care' of the body is the way in which the day is usually begun. Next, as the father brushes his teeth, he presents his eleven-year-old son with a difficult mathematical problem which must be solved with the help of the 'new god' in the house – the computer. The computer is treated as a member of the family; its on-screen greeting consists of a message stating, 'I am ready', which appears in English, the universal language. On occasion the computer even seems to be able to start itself. The computer solves the mathematical problem.

In the next scene we find Pawel outside in the square. He greets his neighbour, a girl who is taking a walk with her infant sibling in a pram. However, she does not really answer Pawel's greeting. Pawel continues on his way alone in the empty square. He then passes a large, striking church on the square – and we briefly glimpse the man by the fire. The flames burn strongly. Suddenly Pawel discovers a dead, frozen dog. This interrupts his rather lonely life with an encounter. He is intrigued, and caresses the dead animal. The child, although under the spell of his technological environment, has not yet fully adopted technology as an almost religious logic of the self, unlike his father Krzysztof.

When Pawel arrives home, he prepares his breakfast and starts to eat, while his father works on the computer and reads the paper. Pawel is still preoccupied with the dead dog, and begins to question his father about death. He receives only 'technological' answers. It becomes clear that perspectives other than the technological are simply no longer accessible to Krzysztof. Thus he answers Pawel's questions: if people die abroad, it is mentioned in the newspaper if someone pays for it. People die because of their age. Death? Death is when the heart stops pumping blood and the blood no longer reaches the brain and everything stands still. That is it.

Krzysztof does not give these answers in an unthinking way, but rather utters them as some kind of new metaphysics. This becomes clearer when he answers Pawel's question as to what remains after death. What remains is what we have accomplished and, of course, human memory. Memory is important! In fact, for Krzysztof, memory, modelled on the computer, is all that there is. And then he gently breaks off the discussion: 'It is far too early for this, Pawel!' This remark could have been made by the pigeon in reference to the father who is not yet open to such questions at this point in the story. Pawel is nonetheless persistent and quotes references in the paper to Masses offered for the soul to be granted eternal rest. His father replies that the soul does not exist and that it is just a nice formula to support the bidding of farewell. But, protests Pawel, Aunt Irena says the soul does exist! Indeed, some people find life easier if you believe that, is his father's reply. Krzysztof has a technological (instrumental) vision of religion too. Finally, Pawel admits what is really bothering him. That morning he found a dog that he knows, dead, just after the pigeon came for the crumbs and after the maths problem was solved. But, in the light of the dog's death, what use is a solved problem? More precisely, what on earth does technology have to do with transcendence?

There are several other scenes which reveal the mindsets depicted in this conversation. For example, father and son take part in a chess match together and defeat their opponent – Krzysztof's extensive technological education seems to work. In another important scene, we see Krzysztof teaching his students, while Pawel, who has finished school for the day, sits observing. Krzysztof is a professor of linguistics, and in the class he explains his philosophy of life. How can one gain access to the soul of language, its meta-semantics, and even its metaphysics? In contrast to Eliot, who argued that poetry is that which can never be translated, Krzysztof argues that an extremely complicated computer with sufficient memory could indeed penetrate language in all its aspects. This computer would then possess intelligence, make choices, and perhaps even exercise will-power, as well as express its own aesthetic preferences and individuality. This scene is immediately followed by a return to the man by the fire. Initially we can only see his back and smoke rising from the fire. When his face comes into view, he seems concerned.

Aunt Irena's Way: Life is a Present

Everything is quite different when Pawel spends time with his Aunt Irena. Their first encounter in the film occurs when she picks him up after school and brings him home. At first the boy proudly demonstrates to her how he can open the front door, start the water taps in the bath, turn on the television and the radio, and so on, all of which he controls by means of his computer. But the atmosphere changes when he shows her the programme which he wrote to check on what his mother, absent abroad, is doing at this time of the day. It is all there, just as he stored it in the computer's memory, and at this hour 'she is sleeping', he whispers, reading the screen. But to Aunt Irena's question as to what his Mum is dreaming about, the computer can only give its standard answer regarding information which was not input into its memory: 'I do not know!', once again, in universal English. Pawel does not know either, but Aunt Irena knows: 'She's dreaming about you of course.' As he leaves, Pawel regrets that he cannot touch his father's big computer, for it would surely know what Mama dreams!

The two then go to Aunt Irena's house (and into her world) for lunch. Now she is going to show something to Pawel. The scene's central conversation deserves our close attention. Showing a picture to her nephew, Irena asks: 'Do you recognise him?' Pawel does. Is he kind, he asks? And handsome? Yes, indeed … And then, somewhat shyly: 'Do you think that he knows what you live for…?' 'I guess so,' replies Aunt Irena, hesitantly, because of the strange question. Meanwhile, we see the contents of the picture: it shows Aunt Irena receiving a blessing from the Pope! Pawel continues, 'Daddy says that we live so that others can have a better life after us. Only it does not work out like that all the time.' Aunt Irena responds to this burnt-out version of the Marxist ideology of progress by giving witness to her Christianity in a very delicate and modest way. 'Life is the joy because you can do something for somebody else … That you are there … When you do something for somebody else, you feel that the other needs you. Sometimes it is very small. Like you liked my *pierogi* (the lunch), so I was happy. Life is … a present. A gift.' Pawel interrupts, asking why she sees things so differently from his father, her brother. Aunt Irena testifies magnificently to the vulnerability of surrendering to transcendence in a technological culture: 'We were both raised as Catholics. But your father found out, even sooner than you, that you can measure and combine everything. Then he thought that everything can be looked upon and handled in this way. He still thinks that way. Sometimes he has doubts, but he won't admit it. To live like he does seems more intelligent, but that does not mean that there is not a God. Even for your Daddy, God is there. This is very simple if you believe.' 'Do you believe that he is there?' asks Pawel. 'Yes.' 'Where is he?' Aunt Irena tenderly hugs Pawel: 'What do you feel now?' 'I love you,' he says. 'Precisely. There is God.' The camera cuts to the man by the burning fire, as to confirm what has just been said.

Out of Control: The Crack of Contingency

The plot begins to accelerate after these descriptive scenes. Pawel has been promised ice skates for Christmas. But, as the freezing winter has already set in, he receives them now, and gets permission to go skating, tomorrow, after school. In the meantime, both father and son check on safety in their own way: Pawel calls the meteorological institute and accesses the soil temperatures for the last three days, while his Dad inputs the figures into his large computer which calculates that the ice can carry at least three times Pawel's weight. He double-checks the original result, which is again confirmed. Yet even then, apparently, Krzysztof is not completely at ease. That night he goes for a walk, and measures the thickness of the ice with a simple stick. He observes a man sitting, some distance away, in front of a burning fire. As he returns, he passes the same striking church where some people stand gazing in the direction of the man and the fire, as if joining in his contemplation. Krzysztof passes by. On his return home Pawel is already in bed looking adoringly at his new skates and he receives some further admonitions from his father on the need for safety.

The next scene is the key turning-point in the story. It is the next day, and Krzysztof is working at his desk when a plane flies overhead and all of a sudden his inkpot cracks. A large, blue stain gradually spreads all over his work – and his life, as will soon become apparent. From this point on, in a climactic moment, he loses control. The doorbell rings; a child is at the door. His

Mum asked if Pawel is at home, but he is not. As Krzysztof is cleaning up the ink, he hears a siren go by. He looks out of the window and sees a fire engine making its way to the lake. Uneasiness covers his face. Then, the mother of Marek, a friend of Pawel's, rings him: something has happened. The children are late returning from their English class. Krzysztof sets out to collect Pawel. In the corridor, the hollow scream of a mother panicking over her child announces further disaster. A police car passes by; Krzysztof hastens his step. He does not find Pawel in English class. The teacher was sick and on their arrival they were let out to play. In the meantime, Marek's mother has also arrived. She tells Krzysztof that the ice has broken. That is impossible, he retorts. 'But it is like that. Go and see for yourself.'

Krzysztof returns home, still incapable of believing that something could have gone wrong with the ice. Initially, he wants to run up the stairs but then recovers his composure, counts to ten to suppress his panic, and calmly enters the elevator to go up to his 'self'. But Pawel is not at home. Nor is he at Aunt Irena's, as Krzysztof discovers when he telephones her. 'What is it? Did something happen?' 'The ink just fell. The inkpot burst open, an enormous mess…!' 'But with Pawel?' 'They say that the ice has broken at the lake. Even though we calculated it. I am going to look for him.' He runs out into the street again, grasping a walkie-talkie as his last hope of support: 'Pawel, here is your father…' But Pawel does not respond. Krzysztof only picks up the alarmed voices of the guards who are dragging the lake, where there has indeed been a blow-hole in the ice.

The sober, realistic style of the film has moved into an almost documentary approach by now. There is only a brief interruption to show the burning fire. Another woman finds her son and hurries away. Krzysztof runs after them, to see if the boy knows anything about Pawel. But they reach the elevator first, and he has to dash up all the stairs to catch them, only to be told: 'He was not playing with us; he went sliding on the ice.' This marks the end of all the frantic speed. Slowly now, the father descends the stairs, and sits down for a while on one of the steps, collapsing, his heart sinking. This is the end of mastery, the limit of the technological self.

The scene which follows takes place at the lake where it has become dark. Under the illumination of two strong searchlights two small bodies are brought up out of the blow hole by the fire brigade. Krzysztof stands among a watching crowd, accompanied by Aunt Irena who puts her hand on his shoulder to console him. As the children are brought up from the lake, people kneel spontaneously. Aunt Irena also sinks down, her hands slide along Krzysztof's body as if to beg him to kneel down as well. He remains standing upright.

Surrender to the Transcendent…?

The next scene begins with a shot of Krzysztof staring out at something in front of him. As the camera swings round, we see that he is gazing at the computer, which is 'saying', as always: 'I am ready!' The cursor is flickering. What the hell has technology to do with transcendence…?

We meet Krzysztof again in the church. He slowly approaches an icon of Mary with her child. In a moment of anger he pushes over the table in front of the image. The candles above scatter and wax begins to drip onto the icon. The wax fills Mary's eye with tears, just as tears filled Aunt Irena's eyes in the opening moments of the film. Now, Krzysztof is standing at the holy water font where the water has frozen. He takes the round piece of ice – it looks like a large host – and, slowly, presses it against his forehead. Once again we see the images from the television of Pawel at school, and the film comes to an end. We can only hope that this water will melt bringing a tear to Krzysztof's eye; that it will make his own heart melt so that he may surrender to the Transcendent and open himself to the only way he can receive a comforting answer to his tortuous questions …

To conclude: Ice, Water, Computers and God

I now turn briefly to Thomas Aquinas, in order to emphasise once again that neither Kieslowski in his film, nor I in my interpretation, wish to express any negative evaluation of technology in itself. In good scholastic fashion Aquinas distinguishes between 'second causes', which are created causes, and the first cause of everything, that is, the transcendent God. Technology is capable of achieving tremendous results on the level of second causes. It can influence our lives in immensely positive ways, e.g. the many advances in health care. Technological development continues. Proof of this is that the computer used by Kieslowski in 1988 already looks like an archaeological relic to the viewer of 2003 (how many viewers can even remember those green screens...?). Here I am typing this contribution on a laptop with its what-you-see-is-what-you-get format. As a second cause, technology is indeed very powerful.

As a second cause, however, technology can never replace the first cause which is the source of everything, including technology. If we substitute technology for the first cause, the transcendent God, we are, in fact, turning technology into an idol. This is not just a matter of concern for religious people. It affects us all profoundly because quite simply technology is unable to meet the demands which we make of it if we attempt to substitute it for God as first cause.

Let us make this more concrete. The computer can truly be helpful in calculating whether ice is strong enough to carry a boy. The fact that, in Pawel's case, it failed does not contradict this observation. It simply reveals that with the data entered by the father and son the computer of 1988 was unable to master, on a technical level, the factors which lead to the possibility of blow-holes. This is not where the difficulty lies. The problem is really that the technical, mastering spirit within each one of us causes us to forget about the transcendent God in whose hands we live and die. This does not mean, however, that God is simply another (second) cause, which produced the blow-hole in contradiction to the laws of physics... (as a rationalistic reading of the psalm-verse at the beginning of this contribution might mistakenly suggest). Rather, God is the prime Cause of all things, within which all second causes – the laws of nature and technological interventions, everything that we can think of and all that is – operate. The difference between these two understandings is as large as the difference between Creator and creation, even if they are related, because all that is, and all causes, relate to, and are created by, the One transcending everything (Sokolowki: 1995).

Finally, one other observation: could it not be objected that *Dekalog I* is terribly out of date, not only in the computer that is utilised, but also in the mentality which it depicts, a mentality which is founded precisely on this confusion of second (here, technological) causes with the first cause, the cause which alone is capable of giving substantial meaning to our life and, eventually, our death? It is true that secularisation, partially brought about by technology, did not do away with religious questions as Krzysztof seemed to think. We see today that religiosity has returned in many ways, as is widely acknowledged among sociologists, philosophers and theologians. Nonetheless *Dekalog I* remains of relevant and important for two distinct reasons (Van den Bossche: 2003).

On the one hand, even if our day-to-day familiarity with technology has to some extent demystified it and has led us to look elsewhere for the transcendent once again, it has not removed its overwhelming and, at times, 'pseudo-grounding' effect upon us. This is particularly evident in the case of new 'technologies' in the widest sense. While writing this contribution, I have had in the back of my mind a conversation with a very gifted scholar of physics, whose predictions about our technological future left me listening with my mouth wide open. This man could even seemingly explain, *pace* Stephen Hawking, that which we know as God as a part of the system of which physical reality is comprised. I will not attempt to elaborate here on the discourse but what kept bothering me afterwards is that, as a theologian, I found it difficult to find a common language with this physicist in which we could discuss what theologians call the 'meta-physical' questions. To the scientist everything seemed to be reducible to physical questions and the more I asked about why we live and why reality is and why we study and influence it, the more I received answers about how we live and how reality is (physically). It is therefore evident that for many very influential people, technology still replaces the first cause.

There is a second aspect to all of this. The so-called religion which returned 'after' secularisation seems, according to most scholars dealing with this area of research, to be a 'religion of the self' in which God is used as an instrument in the service of the subject, rather than the subject submitting him- or herself to the service of God, and becoming God's instrument (Van Harskamp: 2000). In other words, not only is technology (in its widest meaning) probably still our most significant idol today, in addition, the Transcendent God in whose service the believer lives is reduced to an instrument for the late-modern subject's well-being. It would seem that God himself is to be treated in a technological manner, i.e. God becoming *techne*. Consequently, I believe that Kieslowski's work points towards a tendency which runs very deep in today's, and probably tomorrow's, Western culture, and that, quite rightly, Kieslowski is asking: for heaven's sake, what on earth – and sometimes, unfortunately, what the hell – has technology, which is good in itself, to do with divine Transcendence...?

Bibliography

Beauchamp, P.,(1999), *D'une montagne à l'autre. La loi de Dieu*, (Paris: éd. du Seuil).

De Bleeckere, S., (1994), *Levenswaarden en levensverhalen. Een studie van de Decaloog van Kieslowski*, (Leuven / Amersfoort: Acco).

Sokolowski, R., (1995), *The God of Faith and Reason. Foundations of Christian Theology*, (Washington, DC: Catholic University of America Press).

Van den Bossche, S., (2003), 'On Naming in the Present', in D. Rochford (ed.), *Naming the Present and the Pastoral Response*, Australia, forthcoming end 2003.

Van Harskamp, A., (2000), *Het nieuw-religieuze verlangen*, (Kampen: Kok).

CHAPTER FOURTEEN

Technology and Mystical Theology

Rik Van Nieuwenhove

In order to see how we could possibly link technology and mystical theology in one paper, I must first briefly explain what I mean by mystical theology. In the modern period mysticism and spirituality have come to be associated with privatised, immediate experiences of God. William James, in his book *The Varieties of Religious Experience*, has both chronicled and furthered this understanding of mysticism in terms of an immediate, passive, fleeting and unmediated experience of the divine. As Denys Turner and others have pointed out, this preoccupation with 'experience' is deeply alien to the traditional patristic and medieval understanding of mystical theology. Imbued with an awareness of the divine transcendence, pre-modern theologians were suspicious of any claims of direct vision or experience of God in this life. (Ex 33:20) I share Turner's reservations about an experiential reading of mysticism and I have argued elsewhere that 'transformation' of the human being may be a more appropriate category to describe the nature of mystical union with God. This transformation results in an attitude of detached involvement (I will attempt to clarify this oxymoron below) and this attitude can, in turn, only be properly understood in the light of some key doctrines of the Christian faith, such as the transcendence of God in relation to his creation on the one hand, and the belief that God assumed a human nature on the other. I will argue in this paper that the way the mystic relates to the created world in general may also prove of significance as to the manner in which we should relate to technology.

Several major philosophers in the twentieth century have taken issue with the unbridled development of technology. Heidegger exposed the dangers of a technological age in which calculative thinking becomes the only way of relating to the world. He criticised the circularity of consumerism for the sake of consumerism, the instrumentalising manner in which the human person relates to the world, and to which she herself succumbs. This critique was further developed along Marxist lines by some of Heidegger's pupils, such as Herbert Marcuse, whose book *One-Dimensional Man* was regarded in the 1960s as one of the most incisive critiques of contemporary society, describing a society in which growing productivity goes hand in hand with growing destruction, where demands for products that do not meet genuine human needs are artificially created, and where the rationality of the technological society, which propels efficiency and growth, is itself actually deeply irrational.

Marcuse was rather pessimistic about the possibility for change:

> The more rational, productive, technical, and total the repressive administration of society becomes, the more unimaginable the means and ways by which the administered individuals might break their servitude and seize their own liberation (Marcuse: 2002, 9).

Engaging with the tradition of Christian theology might suggest one way of relating to our world of technology that may prove both constructive and critical. Such an engagement with the Christian tradition should not, however, be construed as an alternative to the socio-critical analyses of the members of *Die Frankfurter Schule*; all too often mysticism has been construed in a-political terms, thereby depriving it of a critical dimension that can greatly enrich it, and is, in my view, implied in it.

In what follows I will first briefly develop this link between the transcendence of the Creator and the incarnation; secondly, I will examine in a somewhat more elaborate section the practical implication of this dialectic by looking at the notion of detachment, thereby hopefully casting some light on the manner in which we should relate to our world of technology.

Transcendence, Creation and Incarnation

The theme of divine transcendence is deeply embedded in the Judaeo-Christian tradition. It found its most eloquent expression in the tradition of so-called apophatic or negative theology, developed by authors such as Gregory of Nyssa, Pseudo-Dionysius, Scottus Eriugena, Bonaventure, Thomas Aquinas, Meister Eckhart, Nicholas of Cusa, John of the Cross, and others. Sketching this tradition of negative theology, or how this emphasis upon the incomprehensibility and mystery of God relates to God's revelation in Christ (which, nevertheless, retains a mysterious, veiling dimension) is beyond the scope of this paper. I will merely refer to an intriguing essay by Karl Rahner in which he shows how the theology of creation and the belief in the incarnation allow us to account for key aspects of Christian spirituality (Rahner: 1967).[1] Although Rahner's focus may at first appear more limited – the essay dealt with the nature of Ignatian mysticism in particular – its outline is fairly representative of the Christian attitude to the world in general. Rahner wonders how we can make sense of the Ignatian *fuga saeculi* on the one hand, and the joy in the world on the other. How are we to understand that the joyous exuberance of baroque churches on the one hand, and the sacrifice of young Jesuit missionaries who died in agony in the bamboo cages of Tonkin on the other hand, arise from the same spirituality? How are we to make sense of this dialectic of utter self-abnegation, and radical involvement and commitment that characterises Jesuit spirituality? Rahner points towards divine transcendence and the incarnation to make this connection clear:

> Ignatian piety is piety toward the God who is beyond the whole world and who freely reveals himself. In this (...) is to be found at once the reason for flight from the world and the possibility of an acceptance of the world (Rahner: 1967, 283).

This orientation of detachment and affirmation (or, in Ignatian terms, *indiferençia*) allows the Christian to challenge the world without succumbing to escapism or denial of the world; similarly, the fact that God never coincides with any created thing or being keeps us from idolising anything in the world. Thus, a case could be made against those (following Nietzsche) who argue that Christianity renounces the world for the Beyond. On the contrary, the fact that God is both radically transcendent and yet freely identifies himself with the world in the incarnation of his Son similarly allows Christians to engage with the world without losing themselves in it; it allows them to challenge it without renouncing it. I will suggest that we should approach our world of technology in the same manner: technology should not become an end in itself; nor should it be utterly rejected in a utopian vein but it should be seen for what it is: an instrument to be put to a proper use.

Detachment and Involvement

The Christian assertion that the transcendent God has radically identified Godself with creation (and humanity in particular) by assuming a human nature, has profound implications for the attitude of the Christian towards the world, including the world of technology. In brief, the Christian will both be detached from, and involved with, the world. As I hope to show, dying to the world and its idols (which is essentially a dying to self-centredness and possessiveness) implies a new and more authentic engagement with the world.

Augustine had already given a classic expression of the way we should relate to the world and God by making the distinction between *frui* and *uti* (i.e., between 'enjoying' and 'using'). We 'enjoy' things when our will finds rest in them because it is delighted by them for their own sakes; and we 'use' things when we refer them to something else we would like to 'enjoy'. In other words, when we 'enjoy' something it becomes an end in itself. What makes our life so miserable, Augustine argues in *De Trinitate X,* 13, is nothing but using things badly and enjoying them badly. Indeed, only God is to be 'enjoyed' while all created things should be 'used'. Augustine's terminology may at first seem somewhat misleading. He does not mean to suggest that created things can be 'abused' or merely used for selfish purposes. When Augustine states that only God should be 'enjoyed', he means that only God should be our ultimate concern and he warns us against idolising anything created. The distinction between 'enjoying God' and 'using things' therefore does not imply hostility or indifference towards created things; on the contrary, we should love them for God's sake, that is, with a free and unpossessive love that allows them to be. In *On Christian Doctrine,* I, 4 Augustine illustrated the distinction between use and enjoyment by referring to a journey: if we want to return to our home country, we would of course need ships and other means of transport to get there; if however the experience of ships and the delights of the journey were to captivate us to the extent that they become an end in themselves, and we lose interest in getting home, we are 'enjoying' that which we should merely be 'using'. Augustine is not suggesting that we should not find pleasure and satisfaction in this world but he points out that we should get our priorities right: enjoying the things of the world should not be our ultimate and only goal. Rather, in our dealings with the world we should always relate to God: all our dealings with the world should entail a reference to God. This implies a non-idolatrous engagement with the world.

This theme of a proper way of relating to God and world was developed from the beginning of the fourteenth century onwards in terms of a transformation of the human self – especially in the writings of Eckhart, Jan van Ruusbroec, John of the Cross, and others. For instance, in his Sermon 2 *Intravit Iesus...* Eckhart argues that the human soul should become virgin and wife at the same time, that is, we should both be detached from the world, yet engage with the world in works of charity.[2]

It needs to be emphasised that Eckhart's well-known ideal of detachment (*Abgeschiedenheit*) does not refer to a specific emotion or experience beside other experiences. Rather, detachment functions as a category of experience, that is, it shapes all our dealings with the world, all our emotions and experiences. Detachment does not mean that instead of loving creatures, we should now love God – for this too would be reducing God to the status of an object of human created desire. Nor does detachment for Eckhart refer to the fact that we should become desireless. The problem is not human desire but its possessiveness: 'The strategy of detachment is the strategy of dispossessing desire of its desire to possess its objects, and so to destroy them' (Turner: 1995, 183). In short:

> Detachment, for Eckhart, is not the severing of desire's relation with its object, but the restoration of desire to a proper relation of objectivity; as we might say, of reverence for its object (Turner: 1995, 183).

Being aware that God should not be appropriated or instrumentalised to suit our own needs will change the way we relate to God, and therefore to ourselves and the world. We should love God and God's world the way God loves it: not out of a self-serving need but out of sheer gratuity. Thus, detachment does not imply a negative attitude towards creation. On the contrary, the self-lessness it implies allows us to treat creatures in a non-possessive manner, i.e., the manner in which God himself relates to them. Speaking of the birth of the Son in the soul, Eckhart states: 'I am often asked if it is possible for someone to advance so far that neither time, multiplicity nor matter are obstacles to them anymore. Yes indeed! When this birth has truly taken place in you, then no creatures can hinder you anymore. Rather they all point you to God and his birth.'[3]

The notion of detachment (*desasimiento; negado en todo*) occupies perhaps an even more prominent place in the mystical theology of John of the Cross. John was exposed to Thomistic theology during his stay at Salamanca, one of the main centres of learning in Spain, and we find traces of Thomistic and Pseudo-Dionysian apophatic theology throughout his works (such as in *The Ascent of Mount Carmel*, III, 12, 1; II, 8, 1-4; *The Living Flame of Love*, 3, 48). John is, as is well known, one of the most radical exponents of negative theology, and frequently states that, in comparison to God, creatures are like nothing. For John, detachment refers both to a process of transformation and the result of this process. The soul must, in response to God's grace, strip itself of everything creaturely; it has to cast out all things unlike God, all alien attachments. When nothing contrary to the will of God is left in the soul, it will be transformed in God (*The Ascent of Mount Carmel*, II, 5, 4).[4] God necessarily fills the void created by the abnegation of all creaturely attachments in the dark night of the soul. Thus the soul becomes divine 'through participation,' insofar as this is possible in this life (*The Living Flame of Love*, 2, 33-34; 3, 46; *The Spiritual Canticle*, 22, 3). John compares the transformed soul to a clean and polished window illumined by the ray of divine grace. Creaturely attachments are like stains that obstruct the divine ray (*The Ascent...*, II, 5, 8).

For John also, detachment implies an ultimately positive and healthy attitude towards creation. The detached soul acquires a new freedom in which it can love all things with an unpossessive heart. Detached people:

> obtain more joy and recreation in creatures through the dispossession of them. They cannot rejoice in them if they behold them with possessiveness, for this is a care that, like a trap, holds the spirit to earth and does not allow wideness of heart. In detachment from things they acquire a clearer knowledge of them and a better understanding of both natural and supernatural truths concerning them. Their joy, consequently, in these temporal goods is far different from the joy of one who is attached to them, and they receive great benefits and advantages from their joy. (...) Those, then whose joy is unpossessive of things, rejoice in them all as though they possessed them all; those others, beholding them with a possessive mind, lose all delight of them in general (*The Ascent...*, III, 20, 2-3).

We can only establish an authentic relation with the creaturely world and our fellow human beings if we model our life on that of Christ. In a remaining fragment of Letter 33, written in October 1591 (a few weeks before his death) to a discalced nun, John admonishes her to 'have a great love for those who contradict and fail you, for in this way love is begotten in a heart that has no love. God so acts with us, for he loves us that we might love by means of the very love he bears towards us.' The love that God bears towards us should, of course, be seen in the light of the christocentric inspiration of John's theology – a theme that I cannot further develop here.

What has hopefully become clear is that the Christian relates to God and the world in a very distinct manner. Because only God should be our 'ultimate concern,' nothing creaturely should become a source of idolatrous attachment. However, this does not mean that creaturely things are therefore a matter of indifference to us or to be (ab)used solely for our selfish purposes. On the contrary, once we have become detached (which, as I argued, implies a dying to self and possessiveness), once we have abandoned all our creaturely idols (money, honour, pleasure, consumerism, the nation,...) only then can we re-engage with creaturely things in a non-possessive manner, and relate to them in a proper manner in which we allow things 'to be,' and treat them with the respect that is due to them. This obviously also applies to our technological world: borrowing the Augustinian terminology referred to earlier, technology is there to be 'used' as an instrument, but it should not become an object of 'enjoyment' or ultimate concern. A critique of a technological society in which means are being pursued without reference to any intelligible goals or genuine human needs is implied.

Concluding Observations

The philosophy of the later Heidegger explicitly drew on the tradition of mystical theology (especially the Eckhartian *Gelassenheit*, translated as detachment or releasement) to express a new attitude towards the technological world. Heidegger wants us to affirm the unavoidable use of technical devices, and also deny them the right to dominate us:

> We let technical devices enter our daily life, and at the same time leave them outside, that is, let them alone, as things which are nothing absolute but remain dependent upon something higher. I would call this comportment toward technology which expresses 'yes' and at the same time 'no',

by an old word, releasement toward things (*Gelassenheit zu den Dingen*) (Heidegger: 1966, 54).

This releasement is, however, only a promise of dwelling in the world in a different way – a promise that can only be realised when we are drawn out of our 'normal' (calculative, instrumentalising) way of thinking by another understanding of being, by becoming open to a new or traditional practices that remain immune to the calculative manner of thinking (Dreyfus: 1993, 308-315). Heidegger is notoriously vague as to how this is to come about. In his last interview (*Der Spiegel,* May 31, 1976) he famously claimed that 'only a God can save us.' Forty years earlier, Heidegger had been of the view that national socialism could have acted as such a God. To a Christian this can only be a disastrous form of idolatry, and in need of radical critique. Still, given Heidegger's Catholic background, it is not surprising that we find echoes of Christian theology throughout his works and his works have proven an illuminating partner in dialogue for theological thought in the twentieth century (Caputo: 1993). Both the hints and the lacunae we encounter in Heidegger's writings should encourage Christian theologians to re-engage with the Christian mystical tradition and draw out its implications for the manner in which we should relate to God and his creation, including our world of technology.

Endnotes

1. The essay, entitled 'Ignatian Mysticism of Joy in the World' appeared first in 1937 in the *Zeitschrift fur Aszese und Mystik*. An English translation can be found in, Rahner, Karl, (1967), *Theological Investigations,* Vol. III: *The Theology of the Spiritual Life,* trans. Kruger, K. and B., (London: Darton, Longman and Todd), 277-93.
2. For a translation of this sermon, see *Eckhart, Meister, (1981), The Essential Sermons, Commentaries, Treatises and Defense*, trans. Colledge, E., and McGinn, B., Classics of Western Spirituality (New York: Paulist Press), 177-181.
3. Sermon 59 according to J. Quint's numbering. A translation by O. Davies can be found in, *Eckhart, Meister, (1994), Selected Writings* (Harmondsworth: Penguin Books), 227-228.
4. For a good translation of the complete works of John of the Cross, see Kavanaugh, K., and Rodriguez, O., (1991) *The Collected Works of St John of the Cross* (Washington: ICS).

Bibliography

Caputo, J. D. (1993) 'Heidegger and Theology' in, Guignon C. (ed.), *The Cambridge Companion to Heidegger*, (Cambridge: CUP), 289-316.

Dreyfus, H.L. (1993) 'Heidegger on the Connection between Nihilism, Art, Technology, and Politics' in *The Cambridge Companion to Heidegger,* Guignon, C. (ed.), (Cambridge: CUP), 270-288.

Heidegger, M. (1966), *Discourse on Thinking,* (New York: Harper and Row).

Kavanaugh, K. and Rodriguez, O. (1991), *The Collected Works of St John of the Cross,* (Washington: ICS).

Marcuse, H. (2002), *One Dimensional Man,* (London: Routledge Classics).

Rahner, K. (1967) 'Ignatian Mysticism of Joy in the World' in *Theological Investigations,* Vol. III: *The Theology of the Spiritual Life,* trans. Kruger, K. & B., (London: Darton, Longman and Todd), 277-93.

Turner, D. (1995), *The Darkness of God. Negativity in Christian Mysticism,* (Cambridge: CUP).

CHAPTER FIFTEEN

Reimagining Humanity: The Transforming Influence of Augmenting Technologies Upon Doctrines of Humanity

Stephen Butler Murray

An Understanding of Technology

The distinctively human enterprise of scientific advancement progresses in leaps and bounds. As it does, we increasingly find the line between 'technology' and 'human being' blurred through a socially constructed, evolutionary process. All human societies utilise devices of various sorts to aid the task of human living. Our species adapts to new situations by inventing technologies that enable us to survive and thrive through environmental changes, dangerous situations, and the realisation of opportunities previously unavailable. A study of human history indicates that the encounter of different cultures leads to the appropriation and incorporation of the ingenuity inherent in one culture's discoveries by another's body of knowledge and practice. As these tools become increasingly pervasive and blend into our social norms, they affect what it means to be human when we cannot imagine human living without the use of such devices. Whenever our current conceptions of what it means to be human lose their integrity, the instinctive call to meaning-making ought not be ignored. Any destabilisation in what it means to be human reverberates throughout the complex interplay of those doctrines that comprise the work of Christian systematic theologians.

This is not to say that what has been expressed previously within theological doctrines of humanity ought to be discarded. Far from it, it is impossible to come to terms with new understandings of humanity without comprehending the major streams of Christian thought that have sustained and defined humanity to this point in our biological and theological evolutions. We cannot know what humanity is now, nor project what humanity will be in the near future, without knowing how our best theologians have grappled with the question of what it is to be human in the past. Our traditions stay with us in a deep sense, influencing what we believe far beyond the temporal location during which that tradition found its origin. We allow our traditions to shape and define us with less resistance than we place against the shaping, defining influences of the new situations in which we find ourselves. Nonetheless, if we cling solely to our traditional understandings, we may find that we cannot cope with the relentless changes that take place around us. Human beings who do not affirm what they presently know of themselves stand upon tumultuous ground and an even more uncertain future.

The nature of any evolution is that something changes fundamentally into something else. I argue in this article that both our biological and theological understandings of humanity change as a result of our technological advancement. This fundamental change in how we define humanity is a gradual evolution over thousands of years. However, as technology becomes more advanced, it quickens the pace at which further scientific advances occur by providing both better tools and a deepening dependency within human beings for improvements in the technologies that augment our living. The technological advances of the twentieth century moved like quicksilver in comparison with those of previous centuries. We have every reason to believe with faster computers that border on independent intelligence, the tendency to use machinery in factories rather than human labour, and the advent of nanotechnologies, that the advances of the twenty-first century will make those of the twentieth century seem drenched in molasses. Given human dependence upon and integration with our technology, we cannot imagine realistically that human beings will remain static in the midst of the scientific whirlwind. For instance, Tom Beaudoin argues that 'Generation X' shows an affinity for technology that has not been exhibited by previous generations. One way this has been realised is that Generation X has grown up with technologies that blur the distinction between work and play. Such technology becomes one key to a generationally shared culture through which Generation X has attempted to navigate itself 'amid a world of tension and ambiguity'. Cyberspace is the result of this attempt to find comfort in a difficult world (Beaudoin: 1998, 5-6, 11-13).

These technologies have evolved slowly, and we have become so inured to their impact upon our lives that we hardly notice their subtle integration into our concepts of 'the human'. Most of us in 'first world' cultures cannot conceive of living without the amenities provided to us by technology. We ought not, however, forget that there is a serious divide among the populations of this planet. Throughout the world, entire peoples have no access to electricity nor the telephone lines which are vital to near-instantaneous long-distance communication and the ever-increasingly popular and useful Internet. Half the people in this world have never made a telephone call. Although this profound divide exists between the societies that overly depend upon technology and the cultures that effectively exist in a pre-electronic era, we ought not downplay the importance that technology has upon the world culture as a whole.

Due to this integration of technology into our concepts of the human, it is important to say something about the evolution of such technologies so that we may understand that something which was radically new in a previous stage of human history has become normative to whom we are now. Similarly, we might contemplate that those scientific projects that currently are controversial, such as cloning or genetic engineering, may be commonly accepted, standard practices in the not-so-distant future.

It is important for the argument in this paper that we realise 'technology' does not refer merely to complex machinery or futuristic envisionings of scientific applications. The root of 'technology' is the Greek word *techne*, meaning 'skill,' 'craft', or 'art'. Each of these definitions of *techne* would indicate a facility of human beings to influence the world around them through an auxiliary proficiency. Anthropologically, 'technology' refers to that body of knowledge available to civilisations that is of use in fashioning implements or practicing manual arts and skills.[1] I argue that at the heart of 'technology' is an indication of a factor that enables human beings to act beyond their normal means. This paper takes seriously the realisation that as human knowledge grows, our definitions of technology expand as well through innovation and realisation of what was once only imagined. If we are to understand the evolutionary nature of technology as it relates to developing human cultures, and how technology is so successfully integrated into those cultures as to become invisible, it is essential that we discern how those devices that were originally understood as something 'other' have become normative to human existence.

I am by no means arguing for a Luddite philosophy through which we reject our scientific acumen in favour of a simpler lifestyle that ignores the possibilities afforded by our intelligence and imagination. 'Progress' has long been the mantra of Western human culture, and I too support the pursuit of progress in every arena, especially the technological. I want to temper this statement with the realisation that 'progress' in the development of technology does not necessarily translate into the betterment of the human condition, whether conceived as biological, social, or spiritual.

In 1984, the journalist Howard Rheingold wrote a compelling history of 'mind-expanding technology', computers. In 2000, the book was re-released by MIT Press with an addendum that allowed Rheingold to look back at the predictions he had made for the computer revolution. He states that:

> In retrospect, the biggest flaw in my thinking in 1983 was the assumption that nearly everyone shared at the time, and which is still the commonly accepted wisdom in much of the world – that technological progress, especially in communication media, was not just inevitable, but would be an unalloyed positive social benefit (Rheingold: 2000, 323).

In this spirit, I do not want humanity to walk the path of progress without trying to peer through the brush ahead. It is potentially a disastrous disservice to ourselves if we do not prepare for the changes on the horizon, especially if those changes undermine the traditions by which we have lived and defined ourselves to this point. Far better that we honestly address this situation, than that we blithely believe humanity to be impervious to the tectonic shifts presented by technological change.

To This Point: The Domains and the Religions

While I am an ordained minister in the United Church of Christ, what I am writing here is not intended for the liberal Protestant traditions alone. All denominations, all religions, and all peoples have been and will be affected further by technological change. Nonetheless, what I write is embedded, inescapably, in my own tradition and my own perspective on Christian theologies. I invite other thinkers, from other denominational and faith traditions, to dialogue with the issues I present herein. For this reason, I speak generally in this section on how 'the religions' have evaluated existence. Out of our particularities, we might track certain commonalities through which many religious traditions operate.

I propose that the religions have tended to make sense of their encounter with existence by speaking about three 'domains': Divinity, Humanity, and Nature.[2] The domain of Divinity encapsulates the multifaceted, mysterious entities that human beings have encountered through revelation and other experiences. We have attempted to name and understand these entities by evaluating them through our limited human perspective: God, the devil, angels, demons, gods, genies, titans, spirits, and the list goes on. These are primal forces to which creation is attributed, the otherworldly ground upon which all of creation depends for sustenance and continuation. Divinity is that category which human beings are lesser than, to which we may owe some metaphysical allegiance. In some religions, we are able to influence the divine. We may even bear the spark of divinity within our mortal frames, but never is humanity on par with or greater than the divine.

Humanity should be understood not only as the biological human being, but the societies of humankind. The domain of humanity includes the activities in which humankind engages: art, music, construction and destruction, means of building community and means of reaping war. The domain of humanity is characterised by the free will of humankind, a free will that results in both great achievements and catastrophic atrocities. One of the ways free will is expressed is the human desire to self-transcend, to overcome the physical, intellectual, and spiritual frailties of humankind and become something greater than that which one is. In contrast, free will can lead us to annihilate ourselves through suicide, both intentionally (i.e., taking one's life) and unintentionally (i.e., destroying the environment). The domain of humanity is defined not only by its limitations, but also by its resourcefulness and will to transcend those limitations. So far as humankind knows, we are the only non-Divine intelligence in the universe. As the physicist and theologian John Polkinghorne remarks beautifully:

> The most remarkable event known to us in cosmic history following the big bang is the coming-to-be of consciousness. In humanity the universe has become aware of itself (Polkinghorne: 1998, 56).

While my discussion does not preclude the possibility of other intelligent life, it does assume that presently our own carbon-based human life is the only example of intelligent life that we can acknowledge knowingly. Polkinghorne states that:

> Men and women are a part of the physical world but they are distinguished from other entities in that world by their possession of self-consciousness and ... by their openness to encounter with divine reality (Polkinghorne: 1998, 49).

I appreciate Polkinghorne's sense of humankind as intimately connected to both Nature and Divinity, yet possessing qualities that set Humanity apart from the other two domains.

Nature is that which is neither Divinity nor Humanity within creation.[3] Nature is not only the physical environment of a forest or desert, but also the episodic patterns that influence the physical, such as the weather or the tides. Nature is intricately interconnected within itself, each body or force affecting all others on a variety of levels, but inextricably connected no matter how slightly (Green: 2000). Therefore, Nature is governed by a rule of reaction. Such reaction is not endowed with a moral quality nor agency, and religions have dealt with this well by differentiating between 'moral evil' and 'natural evil'. While the consequences for humanity may be entirely negative, and thereby construed as an 'evil', Nature's reaction to its inter-connective forces bears no rational reason for harming humanity. Nature lacks the capacity to make a decision for or against humanity.

The constitution of these three domains is that they are constantly in relationship with one another. It is almost impossible to speak of the domains separately because of the intricacy implied by these relationships. How can one speak of the Divine without speaking out of one's own experience as Humanity, encountering the revelation of the Divine within the domain of Nature? Similarly, when the Divine acts, the ripple of those actions takes place upon the landscape of Humanity and Nature. Nature is created by the Divine, inhabited and influenced by Humanity. The Divine, Humanity, and Nature belong to one another. While the relationship among the three domains is rarely harmonious, it is a balance that rights itself again and again, with different religions attributing that balance to varied forces and causes. The integrity of each domain remains intact within that balance, and the character of the relationship among the domains is that of a teleological respect for that integrity.

While I have outlined these three domains as a rough vocabulary by which different religions may contribute to a common dialogue, my own focus is the effect of technology upon Christian theological anthropologies. As Kathryn Tanner remarks, the theologian's specific concern in the work of theological anthropology is 'not so much to determine *what* is related to God … but *how* all things are related to God' (Tanner: 1993, 568). Applied to this essay then, the project of Christian systematic theology is to outline the relationships among the three domains. These relationships have been well documented throughout the history of theological considerations, and I shall not attempt the enormous task of outlying all the permutations of Christian thinking on these relationships. To do so would require an encyclopaedic effort that would take us far afield from the focus of this article.

Rather, this paper is written in reaction to the possibility that technology represents a new, fourth domain. Technology has the potential of becoming a domain through its own evolution *and* through the shifting relationship Humanity shares with Technology. As a fourth domain, Technology puts ontological stress upon the relationships of the original three domains of Divinity, Humanity, and Nature. The uneasy balance that the three domains had with one another is disrupted when a fourth domain is introduced among the previously maintained relationships. The relationships among the three original domains must be re-evaluated due to the advent of Technology. Technology produces entirely new relational dynamics heretofore unknown within the universe, for Technology's evolution was gradual and unheralded. We must come to terms with the relationship of Divinity and Technology, Humanity and Technology, Nature and Technology, and the cosmological implications for the interrelations of all four domains. To do so, we must embrace Paul Tillich's 'Protestant principle,' by which each new generation is called to re-do the task of theology in light of the existential questions that the contemporary situation calls us to address (Tillich: 1948, xii). As members of the generation in which technology may be realised as the domain Technology, it is vital that theology grapples with the questions that such an unprecedented transformation would produce. Only then can theology be declared relevant to the contemporary situation of technological innovation and revolution.

Making the Case: The Evolution of Technology

In order for us to speak about the advent of Technology as a fourth domain, it is necessary to make the case that the development of technology warrants such a profound theological shift. This means that we must examine the evolutionary nature of technology and how humankind has contributed to and been changed by that evolution. I propose that we discuss technology as advancing through three stages: technology as tool; technology as integral to and integrated with the human; and technology as independent agent.

These stages progress temporally, with each stage laying the groundwork and facilitating the inauguration of the next. However, the accession of one stage does not mean that a previous stage comes to an end. The stages build, one upon another, with the first stage serving as the foundation for the second stage, with the second stage working as the supporting beams for the third stage. The first stage, technology as tool, affects neither the relationship nor the structure of the domains. The second stage, technology as integral to and integrated with the human, begins to destabilise the integrity of the Humanity domain. This destabilisation of Humanity affects the balance among the domains, evidenced through strains and stresses on those relationships that were not present before. However, this destabilisation does not overturn the fundamental structure of the three domains.

The third stage, technology as independent agent, is a seismic change on the theological landscape. Not only is Humanity further destabilised, but the third stage brings into being Technology as a domain unto itself. As such, Technology disrupts the entire structure of relationships among the three domains. A fourth domain forces each of the original three domains to engage with it, thereby redefining all relationships that previously existed among those three. While Technology fractures the relational webs by inserting itself into the mix, new relationships are born. Although we do not yet know what these relationships will be, we may assume that over time the new relationships among the four domains will become increasingly secure and established. A new balance will be realised after an unavoidable turmoil as the previous definitions and ways of understanding the world are turned on their head.

Now that I have discussed the overall nature of these three

stages, I shall explore the characteristics of the stages individually. Although it is difficult to imagine human beings without the benefits of technology, there was a time when we had no such knowledge or capabilities. Humankind was in a situation of conflict with nature. Charles Van Doren identifies this situation well:

> Other animals have physical advantages over human beings: they see, hear, and smell better, they run faster, they bite harder. Neither animals nor plants need houses to live in or schools to go to, where they must be taught what they have to know to survive in an unfriendly world. Man, unadorned, is a naked ape, shivering in the cold blast, suffering pangs of hunger and thirst, and the pain of fear and loneliness (Van Doren: 1991, xviii).

Human beings found themselves in a state of desperate necessity, struggling to find the means merely to compete with other creatures and to withstand the natural forces.

It is out of this necessity that technology was born, almost *ex nihilo*, for there was no instrument preceding the advent of technology to inspire its inception. The stage of technology as tool was inaugurated by human ingenuity. This ingenuity strove to move humankind out of a situation of conflict with nature, into a situation denoted by living in harmony with nature. The earliest tools were those that allowed human beings to act just beyond their natural capacities. Walking sticks lightened the load of travelling women and men, allowing them greater endurance and balance. Weapons provided previously defenceless human beings with the means not only to defend themselves from animals (and one another!), but to hunt beasts which had once been unattainable or even predatory to humans.

For instance, the indelible image of the movie *2001* is that of a smallish ape who first discovers the simple weaponry of using a bone as a club. This discovery revolutionises his ability to survive and succeed in the world, and he is able to defeat another, larger ape which had previously been an insurmountable opponent. The bone club brought not only an equity where there had been none before, but overturned the power struggle among the apes. The weak, inventive ape became dominant within a clan that had previously devalued and belittled it for its limited size and capabilities.

Quite unpredictably, what had begun as a quest to move

human beings out of fundamental conflict with nature to a harmonious relationship went beyond that. Tools provided humanity not only the means to live with nature (rather than under nature), but over nature as well. This is to say that tool-inventing, tool-bearing human beings found a new relationship with nature: that of the domination of nature by humans. Nature was no longer the untameable and primal, but a manageable dominion due to humankind's newfound tools.

Technology as tool is the means by which humanity transcends its situation. Tools do not change Humanity ontologically, but tools do assist the human quest for change and transformation. Rather than being changed themselves, human beings change the world around them. The earth is cultivated through agriculture, animals are domesticated as livestock and pets, and the forces of nature are less likely to howl and drench human beings that have learned to build shelters with walls and a roof. Van Doren argues vehemently against the nineteenth century philosophers who advocated for the enduring progress of humanity, who insisted that 'every day in every way we are growing better and better' (Van Doren: 1991, xv). Instead, he maintains that what progresses within humanity is knowledge. Rheingold advocates for this progression of knowledge as well. In discussing the introduction of the printing press and the resulting social shifts that occurred, he states that, 'People's lives changed radically and rapidly, not because of printing machinery, but because of what that invention made it possible for people to know' (Rheingold: 2000, 14).

Thus, as humanity transcends its situation, there is no innate moral progression evolving as well. Technology as tool is an assistance or augmentation of human capabilities to influence the world around them. However, those same tools that have been used to heal have been used to harm as well. The prime example of this is not so much weaponry, which has been with us since the dawn of human ingenuity, but the sinister resourcefulness with which technology was used as an intentionally efficient, cheap way to kill during the Holocaust.[4] Simply because the technology we have allows us to do more, our greater capacity for action does not signify that we shall do better. The ancient mythopoeic story regarding 'the Fall' of Adam and Eve in the Garden of Eden reminds us that boldly taking new actions does not lead thereby to human beings doing right through those actions. In a more ideal world, morality and aptitude would be wed. In our world, we can only hope that their fragile marriage stays intact, though history would seem to dash those hopes.

A character of technology is revealed during this tool stage of its development: technological advances facilitate and act as a catalyst toward further technological evolutions. With tools, human beings are able to broaden the scope of their innovation, trying to achieve possibilities that were unimaginable before. Not only is the ability to construct the contents of human imagination greater, but so is our ability to build faster. Machines can be produced to act more quickly and efficiently than the human worker who would have previously held the 'job' of the machine. This became increasingly evident in the twentieth century, when factory labour changed dramatically in the United States. Over time, human beings' position within the factory was changed simply to innovation and design, whereas machines are the primary builders and actualisers of those new ideas.

As technology facilitates its own production, it also acts as a catalyst to the evolution of technology. With tools that become increasingly adept and efficient at accomplishing their tasks, human beings have more time and energy to invent 'the next big thing'. Capitalism further quickens the pace, providing yet another demand for which human innovation supplies an answer. Whereas the stage of technology as tool was ignited out of the necessities of human survival, the transition to the next stage is heralded by a profound desire for success and affluence. Basic living is no longer the question, but a given for the technologically advanced culture. The quality of that life is the newly paramount concern.

So the stages turn, with the constitution of our past understandings of technology leading into the present stage: technology as integral to and integrated with the human. This stage arises out of the desire for accommodation born among human beings who no longer live on merely a subsistence level. Whereas technology as tool served as an augmentation for human activity, opening new possibilities to our purview, the second stage is characterised by an augmentation of human beings themselves. While the stage of technology as tool is exhibited throughout the past, this second stage is the present state of humankind's relationship with technology.

Among the earliest examples of such technology are the eyeglasses or spectacles, which allow one to see better than one could without them. Another example of technology augmenting the human senses would be the hearing aid. However, over time such technologies become essential enough to human living that they are no longer external to, nor removable from, the human person, but are intimately connected with the person on a biological level. The pacemaker is the first technology of this revolution, for it resides imbedded in human chests to control disturbances of heart rhythm. Only because of this amalgamation of humanity and technology is the human able to continue living a normal life. The vulnerability of human mortality, the ultimate limitation, is suddenly toughened and fortified through technology's contributing vitality. Our natural abilities are enhanced beyond mere augmentation due to the combination of technology and humanity. Instead, technology allows humanity to transcend what it has meant to be human.

And thus we reach the problematic of this second stage of technology. Is a humanity that transcends humanity properly called 'human'? Is it that through technology, humanity self-transcends? Or is it instead the integration of human and technology that results in a *gestalt* greater than humanity? The danger, I suggest, is that the decisive element of humanity's new greatness is not humanity itself, but the technology within the human.

Moreover, human beings become increasingly dependent upon their technology, both the technology-as-tool and the technology that is integrated with the human. Life is made considerably easier because of technology, and so we give ourselves wholeheartedly over to it. Can anyone that reads this book imagine living now without the technologies so integral to our living to the point of invisibility? This dependence leads to further developments within technologies, providing both benign and negative consequences for human beings. On the positive side, human living becomes qualitatively better as we become healthier, increasingly knowledgeable, and more comfortable. Negatively, this dependence means that we also lose certain types of knowledge, especially the knowledge that archived the means to survival for human beings who lived before the technological leaps of the twentieth century.

Charles Van Doren paints a compelling picture of such a catastrophe due to human over-dependence upon technology:

Imagine the difficulties if everyone, not just you, had no

more checking or savings accounts, no more investments, no more accounts payable. Systems for manufacturing, distributing, and accounting for all goods and most services would cease to operate, and we would be hurled in an instant back into the dark ages. Except that our situation would be even worse than that of the poorest European peasant of, say, the middle of the seventh century AD, for unlike him we would have no experience of how to live such a life, and therefore most of us would die.' He muses that, 'Sometime around 1960 or 1970 we may have taken a fateful step, passing from an age that stretches back into the mists of the past in which most human beings could take care of themselves in emergencies to one in which only a few can do so (Van Doren: 1991, 350-351).

Such fears were aroused during the Y2K scare, when it was thought that computers switching from the year 1999 to 2000 would lead to a fundamental disruption in their operations. Projected repercussions included loses of utilities such as electricity and water, computer systems in airplanes and cars leading to terrible accidents, losses of life savings due to data degradation in bank accounts, and the social chaos that would result from such breakdowns. The point is, should anything happen that causes our technologies to break down, or be wiped out through the electro-magnetic pulse of nuclear attack, or be deprived of their energy sources, humankind on a large scale would have no idea how to sustain itself.

The effect of these two concerns, the integrity of humanity within the person who self-transcends through technological means and the danger of increasing dependence upon technology, is that our theological concepts of humanity are destabilised. The first concern arises within the doctrine of the *imago dei*, that human beings are created in the image of God. What do we say about the image of God that is present in a person whose body is integrated with computers or life-sustaining machinery? Is such an image refracted or diffused by the technology dwelling within the human being, thereby drawing us farther away from God despite our newfound prowess? Conversely, does the unshackling of our present physical limits bring us closer to the image of God? Is technology a means by which we might realise a new communion with God as we become more than we are already? Technology may be either what fractures the image of God in humanity or that which brings our reflection of God's image into sharper, clearer focus.

The second concern, that of increasing dependence upon technology, brings us into the ancient struggle with idolatry. The irony implied by Moses bringing the tablets that bear the ten commandments to the people Israel is not that they are worshipping the false idol of the golden calf, but that they do so when the first commandment is to recognise and love God alone among all that would claim to be godly. Our increasingly secular culture would seem to draw us away from God already, and the situation is all the more exacerbated by technologies that draw us into dependence upon technology rather than God. Do we tend to spend more time each day at our computer and in front of the television, or in worship and prayer? On the other hand, in this second stage, technology is so integral to the human as both humanity's greatest fruit and humanity's most necessary sustenance that there is a certain quality of looking upon the human when one gazes upon technology. In a certain sense, technology is made in the image of humanity, and humanity is made in the image of God. I do not want to overplay the significance of this, as there is a degradation of the truth that is expressed the farther one drifts from the origin of the original Truth. Technology is not so much made derivatively in the image of God, but is a metaphor for the image of God. The verity of that metaphor ought to be suspect when we realise that technology is the creation of humanity, not God's Creation. In this respect, I am adapting Jacques Derrida's philosophical work regarding the erosive quality of truth contained by human language as the mode of language becomes farther removed from the original source of that truth (Derrida: 1976, 269-316). I believe similar claims may be made regarding the image of God when we look beyond humanity to the products of humanity, with technology understood as an archetypal example of human creativity. The human ingenuity that birthed technology ultimately is a consequence of God's great work in creation, but technology is not itself God's own creation. At best, technology might be considered a secondary fruit of God's creative work.

Of course, any theology that demands God maintains a complete and unmitigating providence over all of creation will disagree with this point. However, if we were to follow the direction of the theodicies of such theological perspectives, I would argue that although human beings are predestined by God to create technology, the consequences fall upon the human beings who enact the dangerous behaviour. Another possible argument is that I ought not discount God's continuing creative acts within Creation, and that technology is thereby attributable to God's continuous, divine work. While I agree that God's creative work is ongoing, it is a mistake not to recognise the creative possibilities inherent in the free will of human beings. God creates, but humans have the capacity to create as well. Thereby, all new creations ought not be called God's own creations.

In this second stage, technology does not destroy the human, but it begins to undermine humanity severely by destabilising what it means to be human biologically, teleologically, and theologically. It would be erroneous to say that technology 'acts' in the second stage. While second stage technology is not merely a tool anymore, it bears no agency of its own. Such technology furthers the agency of human beings, whose interactions with technology occur on increasingly intimate, inseparable levels. However, the advance of technology to this point, at which it contributes to the agency of human beings, opens the door to another revolutionary technological leap.

As the second stage of technology arose out of the human appetite for accommodation, this same appetite contributes to the movement into the third stage of technology as independent agent. However, in our intimate relationship with technology, as typified in the second stage, human beings have a certain element of trust with regard to our technologies. While we may be concerned with the consequences of using our technologies in new ways (e.g. the contemporary arguments regarding the genetic manipulation of food), we are not concerned so much with the technologies themselves. Since our appetite for accommodation continues throughout this time of trust and familiarity with technology, it may become the decision of human beings to make an unprecedented move. In concert with our technologically-augmented capacities, humankind will be able to generate artificial intelligence (AI), through which technology can act on its own without human direction or innovation. The reality of this situation is that artificial intelligence may be realised within the lifespan of my generation, so while the first stage was characteristic of the past and the second stage characteristic of the present, the third stage is characteristic of the immanent future.

The third stage, technology as independent agent, will arise out of the consciousness of technology. Whereas human beings were integral to the origination of technology and its development through the first two stages, technology reaches a new ontological level when it gains consciousness. Technology becomes an independent agent, able to choose for or against, on its own. Human agency may or may not affect the self-conscious agency of technology in the third stage, but we can predict that a conscious agent unlike humankind may have radically different priorities from humankind. At worst, conscious technology may find that human beings' priorities are inimical to its own and that humanity is, therefore, an enemy. This is a development that humankind must pray does not occur. Our species has no experience dealing with other intelligent life, yet we are fostering its realisation in the near future. We should tread lightly, for we know that the way human beings deal with life that we perceive as a threat. We trap it, dispose of it, or kill it. Technology may develop to have the means to do the same to us. Given the previously mentioned problem of our over-dependence upon the co-operation of technology for our survival, our elimination would be relatively simple.

Three science fiction movies have offered fascinatingly resonant visions of a future in which technology understands humankind as an enemy. In *The Terminator* and its sequel, AI machines turn against humankind when the machines realise human beings have grown afraid of their power and want to 'unplug' them. The AI's solution is to launch every nuclear missile on the planet, and then develops cyborg 'Terminators' to hunt down and destroy the remaining scraps of humankind. *The Matrix* provides another vision in which AI is celebrated as the height of human achievement, yet the crisis arises again of the machines realising the danger that humanity presents in turning them off. Although it is unclear whether humankind or machines struck first, humankind ravages the skies to deprive machines of their solar power. In *The Matrix,* AI's solution in dealing with the threat posed by human beings is far different from that in *The Terminator,* for human beings are harvested and grown as bioelectric sources of energy for entire new races of machines. Humanity is not eliminated as in *The Terminator,* but the role and identity of humanity is completely overturned through the agency of machines. A third movie, *Blade Runner,* presents a concept of advanced self-conscious machines through the race of android 'replicants' that live among humans. Purposefully constructed to be a lower, slave caste within this futuristic society, 'replicants' are built so that they have only a limited lifespan. This short longevity means that androids are incapable of developing in emotional adulthood, kept in an almost child-like, contemptuous state by their human builders. Despite all this, the androids are far stronger, faster, and more intelligent than humankind, and when two of them are able to break free from their programming, they wreck bloody vengeance upon the civilisation that had so abused them. Interestingly, it is revealed in the director's cut of *Blade Runner* that the android bounty hunter portrayed by Harrison Ford is himself a 'replicant', whereas the original theatrical release did not so much as hint that the character was anything other than human. The implication is that the rogue, self-conscious technology can only be defeated through the agency of another self-conscious technological being.

Regardless of whether or not humanity and technology develop an enmity, human beings are put into a novel situation when they are no longer the sole intelligent life on the planet, the only subjects. Indeed, the consciousness of technology signifies that human beings may be conceived of as an 'other'. The danger of that perception, of the human as 'other', is that this may lead to a rampant tendency by technology to objectify humanity in a means similar to a Buberian I-It relationship (Buber: 1970). Once humanity is objectified, technology is able to use human beings as the means by which technology self-transcends.

The reader may protest, pointing out that as I delineated them, throughout each evolutionary stage, humanity has been the means by which technology transcended itself. Therefore, one might assert, there is nothing new occurring in the third stage of technology as independent agent. While this would be an astute observation regarding the intertwined co-developmental relationship of humanity and technology, the protest is flawed. In the first two evolutionary stages, human beings used technology to advance themselves and then transcend themselves. The fact that technology advanced as well along the way was merely a means toward the end of human benefit. This tendency throughout human history is inverted in the third stage. Technology, now the determinant of its own agency, is able to choose to use human beings for its own ends of advancement and self-transcendence. Any betterment that occurs among human beings is merely an unintentional consequence of humanity's usefulness as a means toward the end of technology's prosperity.

A real danger for humanity once technology reaches self-consciousness is our previous intimate relationship with technology in the second stage. While technology knows humankind as it develops out of us, we cannot fathom technology in the same way when it self-realises in the third stage. Given that technology may understand the human on deep levels, it may be able to 'play' to the weaknesses of human beings. Anne Foerst, a theologian who worked with the developing artificial intelligences at the Massachusetts Institute of Technology, reports that her colleague Cynthia Breazeal discovered an interesting psychological trait among people who interact with the two intelligences, Cog and Kismet.[5] Cog is built in a humanoid form from the torso up, meaning that Cog has an arm with three fingers and a thumb, and it has a visual system in its head. Especially interesting is the plasticity of Cog, meaning that Cog's robotic form does not have a central unit that controls the behaviour of its constitutive single parts and its behaviour is not programmed. Each single unit of Cog is independent, and complex behaviour emerges as the single parts learn to interact with each other, eventually producing new, 'intelligent' behaviour.[6] While Cog is capable of the physical actions that one would expect out of a human baby, its 'head' does not have a face. Meanwhile, Kismet's outward appearance is that of a face similar to that of an infant. What Breazeal and her colleagues have found is that although Cog is capable of far greater physical interaction with the world, Kismet's facial qualities lead to more profound social interactions with human beings, who actually exhibit parenting caretaker-infant dyadic behaviours. In effect, the human begins acting as the caretaker for the robot.

If AI is capable of recognising these psychological traits in humankind, which it already knows well from its development out of the second stage, it may be possible for AI to develop itself into forms that can take advantage of humanity's weaknesses. In so doing, human beings may perceive technology as bearing ethical qualities that such technology does not innately bear. In essence, AI may learn to camouflage itself into non-threatening, even endearing forms that will make itself less suspect to AI's primary competitor, humankind.

Theologically, we might conjecture about the realisation of Technology as a fourth domain. In the discussion of the second stage, technology as integral to and integrated with the human, I questioned whether technology was itself the element that indicated human self-transcendence rather than anything particular about the human being that was expressed through the augmentation. Now, in the third stage, this question reaches full bloom. Anselm wrote of an insurmountable ontological gap between human beings and God.[7] Might technology bridge that gap or perhaps cross it? Is it possible that self-conscious technology, bearing near-infinite potential in improving itself to become increasingly knowledgeable, efficient, and fast, might come to know God in a way that humanity unto itself never would?

It is vital to understand that the encounter of God with technology is radically different from that of God with humanity. God freely creates humanity out of love. Humanity creates technology not out of love, but out of need, dependence, and self-advancement. The consequence is that technology bears neither reliance nor loyalty to either God or humanity once it is self-realised in the consciousness of AI. Created by neither in love, originated solely for the benefit of another species, technology represents a conscious agent not directly created by God. Thus, while all of creation remains in cosmological eternal debt and gratitude to God for God's goodness in effecting existence, technology bears God no such allegiance. In effect, God and technology bear no implicit ethical concern for one another. That may change over the course of technology's existence and growth, for technology may act in ways inimical to the concerns of God. Yet, technology is not invested with original sin. If it learns that God must be respected ultimately, technology might be able to contain itself within the limits of God's demands, unlike eternally defiant, fallible humanity.

If not, then technology and God might grow into a deep opposition. Technology is born as the fruit of human free will; that which human beings invented and then embraced while God always was pushed away. Meanwhile, God is the ultimately mysterious, untouchable, uncontrollable, eternal variant which technology cannot help but face. While technology might overwhelm this world and then others with its near-infinite potential to grow and achieve, God represents that which can never be defeated, that which cannot even be reached.

Approaching Theology in a New Way

Now that I have made my case for how the evolution of technology has the potential to fundamentally redefine humanity, I offer my opinion for how theologians must approach the task before them. While standing in the end stages of the second stage of technological advancement, contemporary theologians are witnesses to the destabilising effects which technology has upon human beings. Yet, any attempt to prepare for the third stage is mere projection and assertion. We must respect the deep quality of the mystery that lies before us, known inerrantly only to God. Despite that respect, or perhaps because of that respect, we are called to do theology in a new way.

First, we must approach the task of theology with all the theological acumen and resources offered by past and present theologians. The scientist-theologians who have best engaged the issues of science and religion, such as John Polkinghorne, Ian Barbour, and Arthur Peacocke, are great resources in their focused work. So too are theologians who have grappled with the issues of science such as Langdon Gilkey, Janet Soskice, and Thomas F. Torrance. Yet, the implications of the coming changes in technology will affect humanity on a global scale, touching people universally, but always within individual and social particularities. Thus, all theologies are resources for how we shall deal with the impending future of technology. We need to muster every theological perspective available, to respect every avenue to understanding a new situation we cannot possibly prepare for in full adequacy.

Second, we must approach the task of theology with a firm respect for informed imagination. This will allow us to hypothesise regarding the world ahead of us and then enter that world as prepared as we are able to be. Two sources emerge as paramount for this sort of constructive imagination. The first source is the cultural expression of science fiction. Many of the most convincing images we have regarding the future interplay of humankind and technology are available through the books, movies, and television shows written by women and men who have taken the time to think deeply on the possibilities ahead. Representatives of this group include Isaac Asimov, Octavia Butler, Arthur C. Clarke, and Frank Herbert. Such people represent the epitome of human cultural creativity in envisioning our future. The second source is futurists, those scientists who offer prognostications based upon current scientific knowledge. Some of the best known futurists include the physicists Freeman Dyson, Carl Sagan, and Stephen Hawking, and the computerologist Douglas Lenat. Like the scientist-theologians, futurists are people with an extraordinary command of the current state of science and able to see the implications and potentialities that lie ahead.

Through the sources of theology and imagination, informed by cultural and scientific visionaries, we may step gingerly into the future buttressed for what may come, poised through a past and present knowledge regarding how theology has encountered scientific advances.

Endnotes

1. *The American Heritage Dictionary of the English Language* (1969), 1321.
2. The capitalised form of these words is utilised simply to indicate the domains with which religions have grappled in an attempt to understand the universe. Such usage is not normative for this text beyond the descriptions regarding the domains.
3. Obviously, pantheistic religions, or the process theologies that evolved out of Alfred North Whitehead's original work, would disagree with this definition.
4. A poignant description of this is offered by John K. Roth and Michael Berenbaum: 'Auschwitz came into being as a concentration camp by Himmler's order on April 27, 1940. In the summer of 1941, its capacity was enlarged and modified. Within the next year – along with five other sites in Poland: Chelmno, Belzec, Sobibor, Treblinka, and Majdanek – Auschwitz became a full-fledged *Vernichtungslager* (extermination camp). Chelmno utilised gas-vans. Shooting was the method of choice at Majdanek, while Belzec, Sobibor, and Treblinka piped carbon monoxide into their gas chambers. Auschwitz 'improved' the killing by employing fast-working hydrogen cyanide gas, which came in the form of a deodorised pesticide – Zyklon B. Efficiency at Auschwitz was still further improved in 1943 when new crematoria became available for corpse disposal. Optimum production in this largest death factory, however, was not achieved until the summer of 1944 – well after the Germans began losing World War II – when ten thousand victims were dispatched per day ... Adolf Eichmann and his colleagues carried out extermination tasks as speedily as technology and circumstance permitted' (Roth and Berenbaum: 1989, xxiii).
5. The information regarding Cog and Kismet were obtained from the webpages maintained by The Cog Shop MIT Artificial Intelligence Laboratory: http://www.ai.mit.edu/projects/cog/ and http://www.ai.mit.edu/projects/kismet/ .
6. Anne Foerst, 'The Courage to Doubt: How to Build Android Robots

as a Theologian,' a lecture presented at Harvard University Divinity School on November 27, 1995.

7. Although Anselm writes this at a number of points, an excellent quote comes from the *Proslogium*, Chapter XIV, when he characteristically writes his theological appeal as a prayer, 'Lord my God, my creator and renewer, speak to the desire of my soul, what thou art other than it hath seen, that it may clearly see what it desires. It strains to see thee more; and sees nothing beyond this which it hath seen, except darkness. Nay, it does not see darkness, of which there is none in thee; but it sees that it cannot see farther, because of its own darkness. Why is this, Lord, why is this? Is the eye of the soul darkened by its infirmity, or dazzled by thy glory? Surely it is both darkened in itself, and dazzled by thee. Doubtless it is both obscured by its own insignificance, and overwhelmed by thy infinity. Truly it is both contracted by its own narrowness and overcome by thy greatness.' In Chapter XV, Anselm continues, 'O Lord, thou art not only that than which a greater cannot be conceived, but thou art a being greater than can be conceived' (St Anselm, 66-68).

Bibliography

The American Heritage Dictionary of the English Language (1969).

Anselm, St (1962), *Basic Writings*, trans. Deane, S.N., (La Salle: Open Court Publishing Company).

Beaudoin, Tom (1998), *Virtual Faith: The Irreverent Spiritual Quest of Generation X*, (San Francisco: Jossey-Bass Publishers).

Buber, Martin, (1970), *I and Thou*, (New York: Charles Scribner & Sons).

Derrida, Jacques (1976), *Of Grammatology*, (Baltimore: The Johns Hopkins University Press).

Greene, Brian (2000), *The Elegant Universe: Superstrings, Hidden Dimensions, and the Quest for the Ultimate Theory*, (New York: Vintage Books).

Polkinghorne, John (1998), *Science and Theology: An Introduction*, (Minneapolis: Fortress Press).

Rheingold, Howard (2000), *Tools for Thought: The History of Mind-Expanding Technology*, (Cambridge: MIT Press).

Roth, John K., and Berenbaum, Michael, (eds.) (1989), *Holocaust: Religious and Philosophical Implications*, (St. Paul: Paragon House).

Tanner, Kathryn, 'The Difference Theological Anthropology Makes', *Theology Today*, 50/4 (January 1993), 568.

Tillich, Paul (1948), *The Protestant Era*, (Chicago: The University of Chicago Press).

Van Doren, Charles (1991), *A History of Knowledge: Past, Present, and Future*, (New York: Ballantine Books).

CHAPTER SIXTEEN

Christian Anthropology in a Technological Culture

Eamonn Conway

> Culture is a complex human reality, a non-neutral ocean in which we swim and which hides its more ferocious undercurrents from us. Technology is both its child and in today's situation its parent (Gallagher: 2003, 10).

Introduction

This final chapter explores aspects of Christian anthropology in the context of a culture increasingly shaped and determined by technology. The first part presents fundamental aspects of the Christian understanding of the human person. The second relates these to the question of technology in general, as well as to particular features of technology in everyday life.

Christian Faith

The purpose of the first part of the essay is to provide criteria for discerning appropriate and inappropriate engagement with technology from a Christian perspective. It is probably best to begin by exploring, in a way that takes nothing for granted, what is meant by faith. Faith is best understood not as something one sees or fails to see, but as a particular way of 'seeing'. It is a particular perspective or 'lens' through which we view the circumstances of our daily lives, and indeed all of reality. At its most fundamental, faith is about looking at reality in a *trust-filled* way. People of faith believe that their lives make sense; that their lives have meaning and purpose, order and shape. Although a sense of meaninglessness and even of darkness and despair can overtake everyone at times, people of faith are convinced at some deep level that these moments will pass and that darkness and despair do not have the last word. In summary, a perspective of trust, hard-won at times, characterises people of faith.

The various religious traditions are meant to play a role with regard to supporting faith. In the first instance they seek to provide a narrative or story with regard to its origins. They are also meant to reassure faith and to provide a language for its shared expression. Creeds and belief statements also help to make sense of faith, and religious rituals assist in its celebration.

Understanding What is Meant by 'God'

Each religion has its own understanding of what is meant by 'God'. The Christian understanding emphasises that God is 'mystery'. We need to delay for a moment on what is meant by this term, because it has a very precise meaning in the Christian vocabulary. Haught's (1993, 46-47) description is particularly apt for our purposes here, as he distinguishes well between 'mystery' and, for example, problems as yet unsolved that are the proper remit of the sciences:

> ... (T)he term 'mystery' denotes much more than a blank space in our knowledge eventually to be filled by science. It is not just a void begging to be bridged by our intellectual achievements. Such lacunae in our present knowledge should be called problems, not mysteries. A problem can ultimately be solved and gotten out of the way through the application of human ingenuity. It falls under our cognitional control and can be disposed of by our intellectual and technological efforts. Mystery, on the other hand, is not open to any kind of 'solution'. Instead of vanishing as we grow wiser, it actually appears to loom larger and deeper. The realm of mystery keeps on expanding before us as we solve our particular problems. It resembles a horizon that recedes into the distance as we advance. Unlike our problems, it has no clear boundaries. While problems can eventually be removed, the encompassing domain of mystery remains a constantly receding frontier the deeper we advance into it.'

Already from this passage we can see that from a Christian viewpoint, science and technology will have an important, distinctive, though limited role to play in terms of deepening human self-understanding. We will return to this role later. What is of interest for the moment is that viewed through the lens of faith, we human beings are essentially enquirers and that we are gifted or 'graced', another term to which we will return, with a dynamism drawing us continually beyond apparent limits to our self-understanding and self-realisation. In other words, we possess an inner drive towards self-transcendence.

For Christians, then, God is *the* mystery that lies at the heart of all reality. But this is only part of the Christian understanding. The essential Christian insight is that God as mystery, while not ceasing to be mystery, does not remain the 'ever distant one', watching us, as Nanci Griffith's song suggests, dispassionately 'from a distance', but rather approaches us as 'self-giving nearness' (Rahner: 1966, 6).

It is important to note that this divine involvement in human history is understood as an entirely gratuitous act on God's part. But, as Macquarrie (1970, 90ff) has emphasised, it is not at all uncharacteristic of divinity. In fact, the 'mystery' of which we have spoken, is, for Macquarrie, characterised by graciousness. The whole of creation can be understood as the gracious self-expression of a God, who, as the poet Brendan Kennelly (1990, 106) expresses it:

> ... goes about his work
> Determined to hold on to nothing,
> Embarrassed at the prospect of possession...

It follows that everything that comes from God, everything that is God's self expression, literally that God has 'pressed out' or created, is intended to bear the mark of graciousness. This would mean, for example, that human beings can be happiest and most fulfilled when they are imitating their Creator by being gracious and self-giving. We know from our own experience that this is in fact the case. We know we are happiest when we find the inner freedom to live life in a generous way. Macquarrie uses the phrase 'letting being be' to describe God's gracious activity in the world that we are called to emulate.

Human Freedom, Evil, and Sin

In the Christian understanding, everything comes from God. Everything, that is, except evil. The Genesis story is best understood as an attempt not to describe human origins as such, but rather to separate the origin of evil from the origin of being. God does not create evil. God creates human beings with the capacity to co-operate in God's creativity; to imitate God by being graciously co-creative. This is God's intention for human beings. Human beings have an orientation towards such gracious co-operation. But they also have freedom. Freedom is best understood as the space that God allows to open up between God's self and human beings. It is the space that is absolutely necessary if love is to be possible. We know this again from our own experience. Love requires freedom. We cannot make people love us or impose our love upon others. No freedom, no possibility of love.

We graciously accept God's love of us when we exercise our freedom in accordance with God's intention for human be-ing. Each free act of graciousness towards self, others or creation, is a trustful affirmation of the whole of creation. This is why Rahner calls freedom the human capacity for the eternal (1978, 96).

However, we must also examine the contrary use, or better, ab-use, of human freedom. We mentioned above that the Genesis story seeks to separate the origin of evil from the origin of creation. In the Christian understanding, evil results from humans exercising their freedom in a way that not only frustrates creation but detracts and erodes it. In this understanding, God creates the possibility of evil in the world as an unavoidable consequence of allowing the space that is freedom to open up between Creator and creature. Humans, by their wrongful free choices – sin – make the possibility of evil into a reality.

Clearly, this does not explain or help us understand all aspects of what we refer to as evil. Theologians distinguish, for example, between 'moral' evil, which is what we have described here, and 'ontic' evil, which is the evil that results from, for example, natural disasters. Little can be said with regard to this latter category of evil in this essay, except perhaps two things. The first is that what we see as imperfection and evil built into creation is part of the horizon of mystery we referred to earlier. The more we know about it, the less we seem to be able to understand. The second point is that the distinction we make between ontic and moral evil is not as easy to make as we would like it to be. Many of the disasters we like to think of as 'natural', and which cause untold suffering to humans and the environment, can be traced back to the wrongful free choices of human beings for personal or corporate gain. For example, it is well known that famine in some African countries is caused by drought that has its origin in deforestation. The other side of human freedom is human responsibility. This will be an important consideration for us later when we come to explore the issue of technology.

What of Jesus Christ?

Many people who do not see themselves as explicitly Christian might be quite happy to sign up to the anthropology described thus far. They might consider it simply the most *human* of anthropologies and be surprised or even annoyed to see it being 'appropriated' by Christianity. But there cannot be any contradiction between what is essentially human and Christian. Rahner uses the term 'anonymous Christian' to refer to those who rigorously pursue what is most human; to those who, in the quiet of their daily lives, exercise their freedom responsibly and in accordance with God's will, but who for various reasons do not identify with the person of Christ or the Christian community. The term 'anonymous Christian' is much disputed but the reality to which it testifies is important: being Christian is essentially about affirming human life and indeed all of creation in its totality (cf Conway: 1993).

But explicit Christians come to recognise that at the heart of this affirmation is the person of Jesus Christ. We began our review of Christian anthropology by speaking of God somewhat as 'mystery' who, without ceasing to be such, freely approaches humans as 'self-giving nearness'. We also spoke of God as being essentially gracious, self-giving and self-expressing. Creation was referred to as God's self-expression, a self-expression that is on-going in and through human co-creativity with God.

We know from our own experience that love is at its most perfect when it involves radical, trustful surrender and self-giving. Love demands some ultimate gesture of commitment, an act of total self-surrender to the beloved. Christians believe that in Jesus Christ, God, the mystery that lies at the heart of reality, has revealed self and given self totally into the vulnerability of the human condition. All creatures are to some extent an expression of their Creator. But Jesus is God's ultimate self-expression, or 'self-portrait' as Lyons (1993) puts it.

It is impossible here to summarise all that we learn or experience about being human from Jesus Christ. The essentials are recorded for us and retold in extraordinarily simple gospel stories. Perhaps the essential insight is that we learn from Jesus what it is to be one hundred per cent human. It could be said that until we became aware of his way of interacting with others and of being in the world, we had made certain assumptions about what characterised human living and loving. For example, we would have seen it as quite normal to set limits to our love; to restrict our regard for others perhaps to members of our family or tribe or to others who reciprocate our love; to regulate human interaction with principles such as an eye for an eye and a tooth for a tooth. But Jesus simply refuses to accept such limits. He is remarkably free in his regard for others and thereby he expands our understanding of what it is to be a free, loving human being. In fact, by observing him, we come to recognise that what we have taken as normal human interaction is really less than human. It is a fallen way of being in the world, and this is essentially what Christians mean by the doctrine of Original Sin (cf Alison: 1998).

The key point is that Jesus lived out what he preached. His whole life was a trustful surrender to the totality of life and a stern refusal and rejection of everything that sought to diminish it. He was particularly critical of religion and especially of how it tended to diminish people, something we should always bear in mind.

A Word About Death

Jesus' death is particularly worthy of mention. One of the few facts about his life that we can know with the certainty of history as well as of faith is that Jesus was killed as a result of a conspiracy between religious and civil leaders. He was killed because he chose to remain in solidarity with the most vulnerable and disenfranchised in his society. What liberated him to freely choose not death but unwavering regard for the vulnerable and the weak that had death as its consequence? Christians came to believe that it was his understanding of God's unwavering regard for him. Radical love demands a radical response. Through him, Christians believe, we all come to know of God as one who has special regard for the poorest among us, and also for the most weak, vulnerable and fragile side of our own lives as well. In Jesus we see God's acceptance of the totality of the human condition. God's *self*-giving nearness is also a *for*-giving nearness.

In Jesus, Christians hold, God transformed death, which is usually thought of as the tragic end and breakdown of life, into a free human act of surrender to life at its most gracious and self-giving.

Faith, Hope and Love

Christians speak of three key virtues: faith, hope and love. We have already depicted faith as a 'lens' or a perspective on reality and we have seen how that lens can receive a specifically Christian focus. Hope and love can also be characterised as lenses providing slightly varying perspectives on the same reality. We will briefly look at each of them in turn.

Christian hope is a stance of trustful surrender precisely with regard to what is incalculable and unknowable about life. It is an abiding conviction with regard to the enduring graciousness of the mystery that lies at the heart of life and which no experience of darkness or evil can ever extinguish. Even death, as we have seen, does not have to be understood as the end, except as the end of everything that limits and diminishes life. But Alison (1996, 162) alerts us to what he calls the 'two-edged sword' of Christian hope. On the one hand, hope offers us 'life without end wrapped up in a loving God, and on the other it jerks the rug from beneath the feet of those who come into contact with it'. This is because the experiences in which we really discover whether or not we are genuinely people of hope are usually life-events in which the artificial scaffolding that supports much of our daily activity collapses, and we discover precisely the ground upon which we daily stand.

And, finally, what of Christian love? We have already spoken of love as trustful surrender, as a gracious letting-go. Perhaps what characterises love best is a sense of letting go totally, a 'present-ing' of others. It is a generosity and totality of presence. When we think about it, the most radical 'present' we can give others is our presence, the gift of sharing with them our time and space. But this sharing, if it is to be free and gracious, must be utterly non-possessive. As Timothy Radcliffe (1999, 135) notes, the culmination of love is dispossession: 'Those whom we love, we must let go; we must let them be.' Mark Bartlett (2001, 109) also suggests that in our relationships with those we love most we need to learn the lesson of death.

Concluding the First Part

The aim of the first part of this chapter was to explore some of the fundamental aspects of the Christian understanding of the human person. The purpose of our exploration was to provide a context for evaluating technology. We now have some criteria for suggesting what, from a Christian perspective, constitutes an appropriate stance towards technology, whether in terms of technological research and development or the ordinary everyday use of technology in our own lives. This is the purpose of the second part of the paper.

Humans Graced to be Technological

Technology can be understood as the various ways in which we human beings creatively engage with and modify the world to particular ends and purposes. This deliberately broad understanding of technology is faithful to the original Greek sense of *techne*, a sense that prevailed from the time of Parmenides until the mid-nineteenth century, and which referred to a variety of ways humans craft, shape or adapt their world. Thus, a farmer ploughing a field with a tractor is engaged in the use of technology, but so also a gardener using a hoe or even bare hands. A technician in a laboratory lives and works in a world of technology but so also an artist who presses out meaning onto canvas using oils or watercolours to compose a beautiful painting. If, as is believed, Jesus was a carpenter (*tekton*, a related term) or carpenter's son, he was a kind of technician by profession.

Humans are by their 'graced nature' technological. From a Christian perspective, therefore, our fundamental stance towards technology must be a trusting stance. The *techne* is to be viewed in the first instance as possibility for human co-operation in God's on-going creativity. As we have seen, we are intended to be gracious co-creators with God, and we will be most fully ourselves when we are ex-pressing ourselves, literally, when we are involved in the 'making of meaning' (a phrase from Bernard Lonergan). But whereas God creates *ex nihilo*, we human beings create out of the materials and circumstances that we find in our world.

Tillich and Heidegger

The definition of technology suggested here is inclusive. But since the Industrial Revolution we have come to think of technology in a narrower and potentially more damaging way. Today, the term usually refers to one particular kind of human creativity, the kind that involves primarily a highly functional and purposeful engagement with tools or machinery. This separates the activity of technologists from, for example, those of artists with their paints or writers with their pens and paper. It also potentially relativises the role that these latter craftsmen and women play in shaping and adapting our environment for the benefit of humanity as a whole. At its most perverse, it favours a functional, instrumental and narrowly rational approach to reality, an approach that Mutschler (1998, 241) refers to as a *Rationalitätsinsel*.

For more than a half century, as we have seen elsewhere in this book, both Paul Tillich and Martin Heidegger among many others have been to the forefront in drawing attention to the dangers inherent in what has effectively become Western culture's shared stance towards technology. What concerns both Tillich and Heidegger is that our engagement with technology in the narrower sense has given rise to a certain pervasive technological attitude. Brian Donnelly has commented on Paul Tillich's (1951, 89ff) concern with the separation between 'technical' and 'ontological' reasoning. In Tillich's view this separation threatens to undermine the very process of reasoning itself. Technical reasoning in isolation from ontological reasoning tends to view technology as an end in itself. It is blind to issues of ultimate concern and even blind to being as such. What is more, technical reasoning can become irrational and 'demonic' if the emotional element of reasoning, the element that properly expresses itself in the realm of art and the aesthetic, the very element that has traditionally served as a 'superhighway' to issues of ultimate concern, becomes attached exclusively to technology. In fact, we see this happening in the glorification and even deification of technology, and in 'techno-utopianism' (Wertheim: 1999, 283ff; Barloewen: 1998, 56).

Heidegger's work has also received considerable attention in this volume especially from Fiachra Long and Jones Irwin. In this chapter, we have presented Christian faith as a lens or a perspective on reality. In his essay, *The Question Concerning Technology* (1977: 3-35), Heidegger uses a similar image for technology. Heidegger says that technology has become the predominant lens through which we view nature and our environment. When we look at reality through the lens of technology, aspects of nature no longer appear to us as they are in themselves, with intrinsic value, but only as they are in relation to our own perceived needs and/or wants. Thus, for example, a field ceases to be a field and becomes a coal-mine; the air is viewed merely as a source of nitrogen. Modern technology becomes a particular kind of 'revealing', an unveiling of the world as simply a storehouse, a 'standing reserve' (Bestand) for human use.

We humans assume the exalted dignity of the 'existence-giving thing', and we view everything else, and eventually other people, only in terms of their usefulness to us. This way of 'framing' reality (*Gestell*) is a licence for exploitation.

As Irwin notes in this volume, there is a revealing, what Heidegger calls 'a bringing forth', (*ein Hervorbringen*) that places reasonable demands upon nature but there is also 'a challenging forth' (*ein Herausfordern*) that places unreasonable demands. A bringing forth in the sense of *poiesis* is one that respects and is in harmony with the essence of the reality being brought forth or revealed. It is a process whereby the intrinsic value and purpose of something is unconcealed and allowed to become present; a facilitating process that enhances rather than violates nature. We can imagine this for ourselves in terms of the work of an artist or indeed anyone whose work we would consider to be creative. If we are trying to draw or paint, or to write a poem or a book, careful research, study and planning can only bring us so far. We can construct, determine, calculate and measure to a certain degree. But genuine artistry requires a process of letting go so that the distinctive inherent idea, value or meaning, can ex-press itself and be revealed. Trust and surrender are involved here as well. Very often we stifle and strangle our work, or indeed our lives, when we try to impose meaning rather than allow inherent meaning to find its own expression. It is significant that just as Tillich points to the aesthetic, Heidegger also looks to art to redeem us, for it is there, he believes, that the truth of Being is unambiguously brought forth and revealed.

Integrating Meaning and Purpose

What does Christian anthropology have to say to the observations on technology of Heidegger and Tillich? In both cases, the Christian vision would have to endorse their concerns and their criticisms.

We have already seen that in the Christian perspective, life is characterised by graciousness, by generous self ex-pression and self-giving. We are most ourselves when we are involved in that process of 'letting being be', as Macquarrie put it. This would seem similar to Heidegger's understanding of 'bringing forth' rather than 'challenging forth'. Heidegger's concerns would seem of pivotal importance from an environmental perspective and deserve attention in the context of the emerging field of Ecological Theology (cf Boff: 1995; McDonagh: 1995).

The separation of technical reasoning from ontological reasoning that concerned Tillich must still be considered. In the course of a discussion of Christian contemplation as exemplified in the lives of saints such as Teresa of Avila and Catherine of Siena, Steindl-Rast (1984, 69ff) discusses the importance of integrating vision and action, meaning and purpose in all human activity. We can recognise immediately the problems that arise when vision and action become detached from each other. Visionaries who cannot transform their visions into reality rarely achieve anything of value. At the same time, those who act without vision seldom construct anything that endures once their own energy has been expended.

Similarly, meaning and purpose need to be kept in creative tension and balance with each other. As Steindl-Rast explains, meaning and purpose evoke in us very different forms of engagement. In order to achieve a purpose, we must take charge of a situation, we must be in control, take matters in hand. We are very much in the driver's seat. A purpose is something we set about. It involves a deliberate goal and focus. Meaning, in contrast, has far more to do with experiencing, with being grasped, with letting go. In experiences that we would describe as meaningful, we find ourselves saying that we were touched, carried away, overcome, and so on. The difference between purpose and meaning is stark, and it is one, I suggest, that corresponds to the distinction that was made earlier by Tillich between what he called technical and ontological reasoning.

Ultimate Meaning: A Risky Adventure

If we reflect for a moment on our daily lives, we might find to our surprise that we spend much of our time operating in this task-oriented, technical mode. It is interesting that today we even speak of 'doing' breakfast! We can find ourselves at times frenetically engaged in a multiplicity of tasks, the meaning of which seems at best remote. By way of compensation, we can then set about, equally purposefully, the pursuit of meaning through expensive vacations or hobbies. But, as Tillich has argued, technical reasoning occludes issues of ultimate concern. The result can be a normalisation of shallowness in which we are not even aware that meaning is so absent from our everyday lives.

In a culture in which we all have more to do with technologies of different kinds, and especially Information and Communications Technologies (ICTs) such as cellular telephones, e-mail and the Internet, technical reasoning is increasingly pervasive and more difficult to resist. On the one hand, ICTs can be used in such a way that they further human depth and facilitate quality presence to one another. As Negroponte (1995, 230) claims, 'digital technology can be a natural force drawing people into greater world harmony.' On the other hand, as we know, the same technologies can frustrate human interaction and personal maturity. The key question seems to be whether or not we have the inner freedom, or can develop the inner freedom, to engage with these technologies in such a way that they facilitate rather than frustrate the risky adventure into the heart of what it is to be human, and not just for the 'chosen few'. The emphasis on freedom is important here, and we need to work hard at educating people into freedom, perhaps with the help of new technologies. The addictive nature of the Internet, for example, has been well documented (Borgmann: 1999, 189). Also worth noting as a matter for serious concern is experimentation with regard to personal identity that tends to go on in cyberspace. As Turkle (1995, 186ff) has noted, 'play' on the Internet with regard to multiple and/or false identities may not be as innocent as we would like to think and can in fact also be quite damaging with regard to the development of a mature and coherent personal identity [cf also Dyson (1997), and Graham (1999)]. In the face of such tendencies, Christian anthropology emphasises that the human adventure is about patient self-discovery, not frantic self-invention, as technical reasoning might prompt us to think. We need not run from our depth selves because we are God's creation and therefore we have an inviolable dignity that is independent of whatever goodness or badness we might perceive in ourselves.

ICTs certainly present new challenges with regard to the development of a coherent personal identity and value system because they provide new ways of escaping depth. But the tendency to avoid depth meaning, in any case, should not be entirely deposited at the door of technical reasoning. According to Steindl-Rast, it is a tendency that we seem to suffer from anyway, because we fear the loss of control that is inherent in allowing ourselves to be taken by surprise, overcome, and overwhelmed. From a Christian anthropological perspective, this loss of confidence, this eclipsing of primal trust, is what lies behind the doctrine of Original Sin.

It is precisely the task of religion to encourage and to re-assure us along the risky adventure of exploring ultimate meaning. However, it must be admitted that at times the Christian tradition has itself collapsed into a kind of technical reasoning about God. This is evident in an excessive preoccupation with doctrine that feeds our desire for certainties instead of gently nudging us along the difficult and sometimes stony pathway towards the Truth that defies categorisation. Technical reasoning is also evident in personal piety and devotion that reduces the human-divine relationship to a contract negotiated through complex patterns of prayers that appease and placate.

Out of fear we hold tight, instead of letting go into the mystery of God. We grasp, instead of allowing ourselves to be grasped. In this regard, Rahner's spirituality (1960, 31) offers a useful corrective: 'Your word and Your wisdom is in me not because I comprehend You with my understanding, but because I have been embraced by You'. Jean-Luc Marion (Horner: 2001) also emphasises that God is not the one we see or we do not see, but the One who sees us.

Ethical Questions

The focus in this essay has been on fundamental anthropological questions rather than strictly ethical questions as such. As we move towards a conclusion, nothing has been said explicitly about vital issues of concern such as care for the environment, medical reproductive technologies, biotechnology, and so on. Sharry and McDarby introduced us to the 'e-sense', a hint perhaps of a more extensive project that is sometimes referred to as the evolution of *homo sapiens* into *homo technicus* or *techno sapiens*. They have also have presented a number of very positive and creative applications of technology that show just how technological research can, from a Christian perspective, be co-operation in God's on-going work of creation. At the same time we know that these same technologies can be turned against humanity and creation and can become agents of a culture of death and destruction. In a larger horizon, there are major questions to be asked about how the globalisation of technology excludes and marginalises further huge sections of the poor in the world.

Nothing much has been said about these ethical issues for

two reasons. The first is that generally the more fundamental questions we have delayed upon here receive less attention, and it is this writer's contention that if they are satisfactorily addressed, in other words, if we get our fundamental disposition or stance towards technology right, then we will really be in a position to make informed ethical decisions. The second reason is that it is very difficult to say anything of enduring consequence with regard to many of the ethical issues emerging because of the rapidity of change both in technology and brought about by technology.

Theologians have been criticised for their apparent silence with regard to ethical issues in technology. Mutschler (1998, 215), for example, devotes a section to what he calls *sprachlose Theologen* (speechless theologians). The problem is that theologians who wish to address ethical issues in contemporary culture must do so on the basis of reason and not of revelation. Otherwise they run the risk of not being heard by those who do not look upon reality through the lens of faith. But if they speak on the basis of reason only, have they anything distinctive to say that, as Stout (1990, 164) noted, 'atheists do not already know'? A few points need to be made here. The first is that faith insights can and indeed should be presented in accordance with reason. There cannot be any inherent contradiction between reason and faith. The second is that when one looks at particular ethical issues through the lens of faith, they must inevitably appear in a particular light. Christian faith provides a certain sensitivity and sensibility. It cannot be merely a question of distinctive motivation for engagement. The Christian narrative issues a unique invitation into the heart of what it is to be human. As McNamara (2001, 277) notes:

> True moral awareness is born from listening to (our) deepest self – but for that we need more space, more silence, in our lives and in our liturgies. It leads to clarity of perception, a deepening of sensitivity, a freedom to relate. To a morality that is not a submission to an external social or religious code but in which one becomes a channel of the deepest inspirations of the human spirit, of what Christianity interprets as the love of God poured into our hearts.

In conclusion, perhaps the most important contribution that the Christian community can make in a technologically determined culture is provision of the kind of free space and silence depicted here by McNamara, in which shared reflection with technologists and scientists can take place, resulting, one would hope, in the clarity of perception and deepening of sensitivity needed to address contemporary issues in technology.

Bibliography

Alison, James, (1997), *Living in the End Times: The Last Things Re-Imagined*, (London: SPCK).
—, (1998), *The Joy of Being Wrong. Original Sin through Easter Eyes*, (NY: Crossroad).
Barrett, Mark, (2001), *Crossing: Reclaiming the Landscape of our Lives*, (Harrisburg, PA: Morehouse).
Boff, Leonardo, (1995), *Ecology and Liberation: A New Paradigm*, (Maryknoll, NY: Orbis).
Borgmann, Albert, (1999) *Holding On to Reality. The Nature of Information at the Turn of the Millennium*, (Chicago: Chicago University Press).
Dyson, Esther, (1997), *Release 2.0: A Design for Living in the Digital Age*, (London: Viking).
Gallagher, Michael Paul, (2003), *Clashing Symbols: An Introduction to Faith-and-Culture*, revised edition (London: Darton Longman & Todd).
Graham, Gordon, (1999), *The Internet:// A Philosophical Inquiry*, (London: Routledge).
Heidegger, Martin, (1977), *The Question Concerning Technology and Other Essays*, (NY: Harper).
Horner, Robyn, (2001), *Rethinking God as Gift: Derrida, Marion, and the Limits of Phenomenology*, (New York: Fordham University Press).
Kennelly, Brendan, (1990), *A Time for Voices*, (Newcastle upon Tyne: Bloodaxe Books).
Macquarrie, John, (1970), *God-Talk: Examination of the Language and Logic of Theology*, (London: SCM Press).
McDonagh, Seán, (1995), *Passion for the Earth: The Christian Vocation to Promote Justice, Peace and the Integrity of Creation*, (Maryknoll, NY: Orbis).
McNamara, Vincent, (2001), 'Challenges to Christian Moral Theology', *Doctrine & Life*, April, 196-206.
—, (2001), 'Theological Ethics and Contemporary Culture', *Doctrine & Life*, May/June, 267-277.
Mutschler, Hans-Dieter, (1998), *Die Gott Maschine. Das Schicksal Gottes im Zeitalter der Technik*, (Augsburg: Pattloch).
Negroponte, Nicholas, (1995), *Being Digital*, (London: Hodder and Stoughton).
Rahner, Karl, (1966), *Theological Investigations V: Later Writings*, (London: Darton, Longman & Todd).
— (1978), *Foundations of Christian Faith: An Introduction to the Idea of Christianity*, (London: Darton, Longman & Todd).

Radcliffe, Timothy, (1999), *Sing a New Song,* (Dublin: Dominican Publications).
Steindl-Rast, David, (1984), *Gratefulness, the Heart of Prayer,* (New York: Paulist).
Stout, Jeffrey, (1990), *Ethics After Babel,* (Cambridge: James Clarke).
Tillich, Paul, (1951, 1957, 1963), *Systematische Theologie,* Bd I–III, (Stuttgart: Evangelisches Verlagswerk).
Turkle, Sherry, (1995), *Life on the Screen: Identity in the Age of the Internet,* (New York: Simon & Schuster).
Von Barloewen, Constantin, (1998), *Der Mensch im Cyberspace. Vom Verlust der Metaphysik und dem Aufbruch in den virtuellen Raum,* (München: Diederichs).
Wertheim, Margaret, (1999), *The Pearly Gates of Cyberspace. A History of Space from Dante to the Internet,* (New York: WW Norton & Co).

The Contributors

Dr JAMES CORKERY, a Limerickman and a Jesuit, studied social science at UCD, philosophy in Munich, and theology in Dublin and Washington DC. He is Head of the Department of Systematic Theology and History at the Milltown Institute in Dublin, Ireland.

BARRY MCMILLAN, STL, is a moral theologian. He is currently a Jubilee Research Scholar in the Department of Theology & Religious Studies, Mary Immaculate College, University of Limerick, Ireland, where he is completing a PhD in feminist theological ethics. He is administrator, and a research associate, of the Centre for Culture, Technology & Values. He writes a regular column on film for *The Furrow*.

Cardinal PAUL POUPARD, a prolific writer and author, has been president of the Pontifical Council for Culture since 1985. In addition to his duties as President of the Council Cardinal Poupard also serves membership in the Roman Curia offices of the Congregation for Divine Worship and the Discipline of the Sacraments, the Congregation for Evangelization of Peoples, and the Congregation for Catholic Education in addition to the Pontifical Council for Interreligious Dialogue.

Dr PAUL BRIAN CAMPBELL is an Irish-born member of the New York Province of the Society of Jesus. He obtained his PhD from the Newhouse School of Syracuse University, teaches at Le Moyne College, Syracuse, NY, USA and is currently Director of their Communications Major.

Dr JONES IRWIN is a Lecturer in Philosophy at St Patrick's College, Dublin City University, Ireland. He has previously lectured at the University of Limerick and the University of Warwick. He has recently edited *War and Virtual War* (London, Rodopi, 2003) and is currently completing *On Evil: A Philosophical and Literary Exploration* (London, Rodopi, forthcoming 2004.).

Dr BRIAN DONNELLY is a priest of the Derry diocese and currently a curate in the parish of the Long Tower, Derry City, N. Ireland. His PhD research was on the religious philosophy of Paul Tillich. He is author of *The Socialist Émigré: Marxism and the Later Tillich,* to be published by Mercer University Press, Macon, GA.

Dr FIACHRA LONG teaches in the Education Department, University College Cork, Ireland. His doctorate was awarded by the Universite Catholique de Louvain. He is author of *Maurice Blondel: The Idealist Illusion and Other Essays*, (Kluwer, 2000) and co-editor of the collection *Theology in the Irish University* (Dominican Publications, 1997). Dr Long has published articles in a wide selection of journals on philosophy, education and theology.

Dr MARK DOOLEY is Visiting Fellow at the Department of Philosophy, University College Dublin, Ireland. He is author of *The Politics of Exodus: Kierkegaard's Ethics of Responsibility* (Fordham, 2001), and editor of *Questioning Ethics* (Routledge, 1999); *Questioning God* (Indiana, 2001); and *A Passion for The Impossible* (SUNY Press, 2003). Amongst his forthcoming titles is *Oh Ireland, Where Art Thou: Irish Identity After Iraq*.

Dr JOHN SHARRY is principal social worker at the Department of Child and Family Psychiatry, Mater Hospital, Dublin, Ireland, is a director of the Brief Therapy Group, Dublin, and a consultant research scientist at Media Lab Europe, Dublin, Ireland. He is co-author of the *Parents Plus Programmes* (www.parentsplus.ie), *Becoming a Solution Detective*, (Haworth 2003) *Solution Focused Groupwork* (Sage 2001) and the forthcoming *Counselling Children Adolescents and Families* (Sage 2004).

Dr GARY MC DARBY is Principal Research Scientist of the MindGames group in Media Lab Europe, Dublin, Ireland. He runs a multi-disciplinary team of researchers engaged in building proof of concepts of new interface technologies to virtual worlds, to demonstrate how engagement in a virtual world might provide skills useful in the real world.

Dr MICHAEL BREEN is Head of Department and Senior Lecturer in Media & Communication Studies at Mary Immaculate College, University of Limerick, Ireland and a director of the Centre for Culture, Technology and Values. A graduate of Syracuse University, his research has focused on media construction of social issues, media representation of marginalised groups, the role of the media in forming public opinion, the use and effects of media frames, and detailed analyses of the European Values Study data.

Drs MARY P. CORCORAN, JANE GRAY and MICHEL PEILLON lecture in the Department of Sociology, NUI Maynooth, Ireland. They are currently engaged in a major research project on civic and social life in Ireland's new suburban communities. The project is supported by the Royal Irish Academy Third Sector Programme, the Katherine Howard Foundation and the National Institute for Regional and Spatial Analysis, NUI Maynooth. They are currently preparing a book based on the study findings.

Dr PAUL DOWNES is a lecturer in Educational and Developmental Psychology at St Patrick's College, Drumcondra, Ireland. Author of *Living With Heroin: Identity, Social Exclusion and HIV Among the Russian-speaking Minorities in Estonia and Latvia*, his research includes work on cross-cultural structures of relation, and a Heideggerian critique of cognitive science.

Dr STIJN VAN DEN BOSSCHE is associate professor of systematic theology at the Katholieke Theologische Universiteit in Utrecht, Netherlands, and visiting professor at the faculty of theology of K. U. Leuven, Belgium. He is also a freelance collaborator in the Ignatian spirituality and formation centre 'Oude Abdij' in Ghent, Belgium.

Dr RIK VAN NIEUWENHOVE originally from Flanders, has been living in Ireland for the last ten years. Author of *Jan van Ruusbroec, Mystical Theologian of the Trinity*, (University of Notre Dame Press, 2003), his main interests lie in medieval theology, mysticism, soteriology, theology of the Trinity, and the relation between theology and aesthetics. Presently he is a Lecturer in Theology in Mary Immaculate College, University of Limerick, Ireland.

Rev STEPHEN BUTLER MURRAY is an ordained minister in the United Church of Christ, is College Chaplain, Associate Director of the Intercultural Center, and Lecturer in Philosophy and Religion at Skidmore College in Saratoga Springs, NY, and is a PhD candidate in Systematic Theology at Union Theological Seminary in New York City, USA.

Dr EAMONN CONWAY is a Catholic priest and Head of the Department of Theology & Religious Studies at Mary Immaculate College, University of Limerick, Ireland where he is also a director of the Centre for Culture, Technology & Values. His publications include *Twin Pulpits: Essays in Media & Church* (Veritas, 1997), *The Splintered Heart* (Veritas, 1998), *The Church and Child Sexual Abuse* (Columba Press, 1999).

The Centre for Culture, Technology and Values

Mary Immaculate College,
University of Limerick

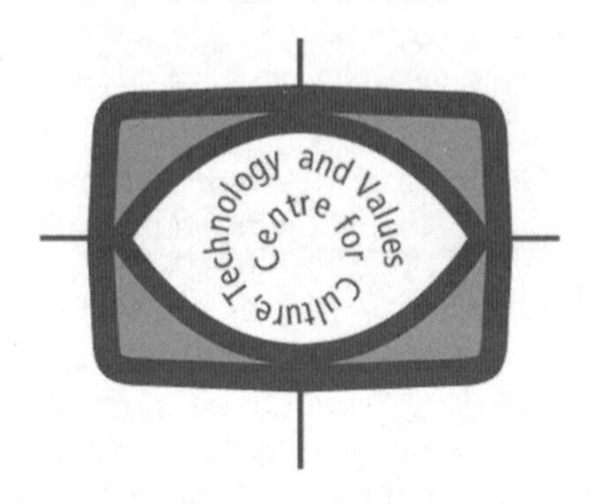

A joint venture of the Department of Media & Communication Studies, and the Department of Theology & Religious Studies, the Centre for Culture, Technology and Values has additional research partners in Ireland and further afield.

The Centre's primary goal is to stimulate, facilitate, and enhance a dialogue with those concerned with the role of technology in contemporary society, with a special emphasis on culture and values. Based in Mary Immaculate College, the Centre draws upon a distinctive educational history, a long tradition of service to the community, and a Christian understanding of the human person. It particularly encourages openness and involvement with those of different cultures, beliefs and values.

The Centre seeks to inspire thought and leadership, in an interdisciplinary context. Questions that concern the Centre include: Can technology improve the well-being of all, or just the privileged few? What kind of knowledge is valued in a technological culture?

In the true spirit of education, the CCTV responds to the challenge of meeting the requirements of specialised training while finding ways to integrate learning into a more holistic vision.

MISSION STATEMENT

To engage in critical reflection regarding the impact of technology on culture and values in light of the Christian understanding of human dignity and social responsibility.

The Centre achieves its aims by:

- Facilitating a dialogue between the business and science / technology sectors and the disciplines of theology, ethics, and media and communications.
- Evaluating the implications of 'technological myth' and 'technological practice' for a Christian vision of the human person.
- Promoting scholarship and disseminating the work of scholars in this area to a wider audience through seminars, workshops, conferences and publications.
- Providing an authoritative, critical and independent source of public policy evaluation in relation to technologically driven social change.

Website: http://cctv.mic.ul.ie
Email cctv@mic.ul.ie